I0467972

NUREG-1875, Vol. 1

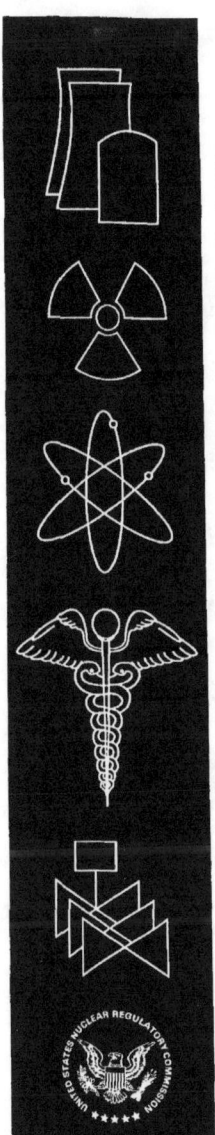

Safety Evaluation Report
Related to the License Renewal of Oyster Creek Generating Station

Docket No. 50-219

AmerGen Energy Company, LLC

Manuscript Completed: March 2007
Date Published: April 2007

**Division of License Renewal
Office of Nuclear Reactor Regulation
U.S. Nuclear Regulatory Commission
Washington, DC 20555-0001**

ABSTRACT

This safety evaluation report (SER) documents the technical review of the Oyster Creek Generating Station (OCGS) license renewal application (LRA) by the staff of the United States (US) Nuclear Regulatory Commission (NRC) (the staff). By letter dated July 22, 2005, AmerGen Energy Company, LLC submitted the LRA for OCGS in accordance with Title 10, Part 54, of the *Code of Federal Regulations* (10 CFR Part 54). AmerGen Energy Company, LLC requests renewal of the operating license for OCGS (Facility Operating License Number DPR-16), for a period of 20 years beyond the current expiration date of midnight April 9, 2009.

OCGS is located in Lacey Township, Ocean County, New Jersey, approximately two miles south of the community of Forked River, two miles inland from the shore of Barnegat Bay, and nine miles south of Toms River, New Jersey. The NRC issued the OCGS construction permit on December 15, 1964, the OCGS provisional operating license on April 9, 1969, and the OCGS operating license on July 2, 1991. OCGS is a single unit facility with a single-cycle, forced-circulation boiling water reactor (BWR)-2 and a Mark 1 containment. The nuclear steam supply system was furnished by General Electric and the balance of the plant was originally designed and constructed by Burns & Roe. OCGS licensed power output is 1930 megawatt thermal with a gross electrical output of approximately 619 megawatt electric.

This SER presents the status of the staff's review of information submitted through February 15, 2007, the cutoff date for consideration in the SER. The staff identified open items that were resolved before the staff made a final determination on the application. SER Section 1.5 summarizes these items and their resolution. Section 6.0 provides the staff's final conclusion on the review of the OCGS LRA.

TABLE OF CONTENTS

Appendices

Tables

ABBREVIATIONS

AAC	alternate AC
ACAD	atmospheric containment air dilution system
ACI	American Concrete Institute
ACRS	Advisory Committee on Reactor Safeguards
ACSR	aluminum conductor steel reinforced
ADAMS	Agency Document Access Management System
ADS	automatic depressurization system
AERM	aging effect requiring management
AFU	air filtration unit
AmerGen	AmerGen Energy Company, LLC
AMP	aging management program
AMR	aging management review
APCSB	Auxiliary and Power Conversion Systems Branch
API	American Petroleum Institute
ART	adjusted reference temperature
ASA	American Standards Association
ASME	American Society of Mechanical Engineers
ASTM	American Society for Testing and Materials
ATWS	anticipated transient without scram
BOP	balance of plant
BTP	Branch Technical Position
BWROG	BWR Owner's Group
BWR	boiling water reactor
BWRVIP	Boiling Water Reactor Vessel and Internals Project
CAP	corrective action program
CASS	cast austenitic stainless steel
CDF	core damage frequency
CFR	Code of Federal Regulations
CI	confirmatory item
CIS	containment inerting system
CIV	containment isolation valve
CLB	current licensing basis
CMAA	Crane Manufactures Association of America
CRD	control rod drive
CRL	component record list
CS	core spray
CST	condensate storage tank
CUF	cumulative usage factor
CVB	containment vacuum breaker
CWS	circulating water system
DBA	design basis accident
DBD	design basis document

DBE	design basis event
DC	direct current
DFED	drywell floor and equipment drains
DG	diesel generator
DWST	demineralized water storage tank
ECCS	emergency core cooling systems
ECP	electrochemical corrosion potential or electrochemical potential
ECT	eddy current testing
EDG	emergency diesel generator
EDGCW	emergency diesel generator cooling water
EFPY	effective full-power years
EMA	equivalent margin analysis
EMRV	electromatic relief valve
EPRI	Electric Power Research Institute
EPU	extended power uprate
EQ	environmental qualification
ESF	engineered safety feature
ESW	emergency service water
F	Fahrenheit
FAC	flow-accelerated corrosion
F_{en}	environmental fatigue factor
FFW	final feedwater facility
FHAR	Fire Hazards Analysis Report
FP	fire protection
FRCT	Forked River Combustion Turbines
FS	feedwater system
FSSD	fire safe shutdown
FWH	feedwater heater
GALL	Generic Aging Lessons Learned
GDC	general design criteria or general design criterion
GE	General Electric
GEIS	Generic Environmental Impact Statement
GL	generic letter
GPUN	General Public Utilities Nuclear Corporation
GSI	generic safety issue
HELB	high-energy line break
HEPA	high efficiency particulate air
HP	high pressure
HPCI	high pressure coolant injection (system)
HVAC	heating, ventilation, and air conditioning
HVS	hardened vent system
HWC	hydrogen water chemistry
HX	heat exchanger
I&C	instrumentation and controls

IASCC	irradiation assisted stress corrosion cracking
ICS	isolation condenser system
ID	inside diameter or identification
IGSCC	intergranular stress corrosion cracking
ILRT	integrated leak rate test
IN	information notice
INPO	Institute of Nuclear Power Operations
IPA	integrated plant assessment
IPE	individual plant examination
IRM	intermediate range monitoring
ISG	interim staff guidance
ISI	inservice inspection
ISP	integrated surveillance program
ITS	important to safety
KIP	1000 lb; or 1 kilo-pound
ksi	one KIP per square inch, 1000 psi
kV	kilovolt
LBB	leak-before-break
LER	licensee event report
LLRT	local leak rate test
LOCA	loss of coolant accident
LOOP	loss of offsite power
LPRM	local power range monitor
LR	license renewal
LRA	license renewal application
MCC	motor control center
MEL	master equipment list
Met Tower	Meteorological Tower
MFED	miscellaneous floor and equipment drain
MFL	magnetic flux leakage
MG	motor generator
MGAS	main generator and auxiliary system
MIC	microbiologically influenced corrosion
MSIV	main steam isolation valve
MSS	main steam system
MTAS	main turbine and auxiliary systems
MUD	makeup demineralizer
NDE	nondestructive examination
NEI	Nuclear Energy Institute
NESC	Nuclear Electrical Safety Code
NFPA	National Fire Protection Association
NITS	not important to safety
NMMS	noble metals monitoring system
NPS	nominal pipe size
NRC	U.S. Nuclear Regulatory Commission

NSR	nonsafety-related
NUREG	U.S. Nuclear Regulatory Commission Regulatory Guide
OCCW	open-cycle cooling water
OCGS	Oyster Creek Generating Station
ODSCC	outside-diameter stress-corrosion cracking
OI	open item
P&ID	piping and instrumentation diagram
PASS	post accident sampling system
PBD	program basis document
PCIS	primary containment isolation system
PDI	performance demonstration initiative
PM	preventive maintenance
PORC	Plant Operations Review Committee
PP	position paper
PT	penetrant testing
P-T	pressure-temperature limit curves
PTFE	polytetrafluoroethylene
PTS	pressurized thermal shock
PUAR	plant-unique analyses report
PWR	pressurized water reactor
PWSCC	primary water stress-corrosion cracking
RAI	request for additional information
RBCCW	reactor building closed cooling water
RBVS	reactor building ventilation system
RCPB	reactor coolant pressure boundary
RCS	reactor coolant system
RDODS	roof drains and overboard discharge system
RFED	reactor building floor and equipment drains
RFP	reactor feed pump
RG	regulatory guide
RHCS	reactor head cooling system
RHR	residual heat removal (system)
ROPS	reactor overfill protection system
RPS	reactor protection system
RPT	recirculation pump trip
RPV	reactor pressure vessel
RT_{NDT}	reference temperature nil ductility transition
RVI	reactor vessel internals
RWCU	reactor water cleanup system
RWSS	reactor water sample system
SBLC	standby liquid control
SBO	station blackout
SC	structure and component
SCC	stress-corrosion cracking
SCS	shutdown cooling system

SE	safety evaluation
SEN	significant event notification
SEP	Systemic Evaluation Program
SER	safety evaluation report
SFPCS	spent fuel pool cooling system
SGTS	standby gas treatment system
SHE	standard hydrogen electrode
SI	Structural Integrity Associates, Inc.
SLCS	standby liquid control system
S_{mc}	stress intensity
SOC	statement of consideration
SR	safety-related
SP	specification
SR	safety-related
SRP	Standard Review Plan
SRP-LR	Standard Review Plan for Review of License Renewal Applications for Nuclear Power Plants
SS	stainless steel
SSC	system, structure, and component
SV	solenoid valve
SWS	service water system
t	thickness
TBCCW	turbine building closed cooling water
TDR	time domain reflectometry
TIP	traveling in-core probe
TLAA	time-limited aging analysis
TOC	total organic carbon
TR	topical report
TS	technical specification
UFSAR	Updated Final Safety Analysis Report
USAS	United States of America Standard
USE	upper-shelf energy
UT	ultrasonic testing
VFLD	vessel flange leak detection
VT	visual examination

SECTION 1

INTRODUCTION AND GENERAL DISCUSSION

1.1 Introduction

This document is a safety evaluation report (SER) on the license renewal application (LRA) for Oyster Creek Generating Station (OCGS), as filed by AmerGen Energy Company, LLC (AmerGen or the applicant). By letter dated July 22, 2005, AmerGen submitted its application to the U.S. Nuclear Regulatory Commission (NRC) for renewal of the OCGS operating license for an additional 20 years. The NRC staff (the staff) prepared this report, which summarizes the results of its safety review of the LRA for compliance with the requirements of Title 10, Part 54, of the *Code of Federal Regulations* (10 CFR Part 54), "Requirements for Renewal of Operating Licenses for Nuclear Power Plants." The NRC license renewal project manager for the OCGS license renewal review is Donnie J. Ashley. Mr. Ashley can be contacted by telephone at 301-415-3191 or by electronic mail at dja1@nrc.gov. Alternatively, written correspondence may be sent to the following address:

License Renewal and Environmental Impacts Program
U.S. Nuclear Regulatory Commission
Washington, D.C. 20555-0001
Attention: Donnie J. Ashley, Mail Stop O-11F1

In its July 22, 2005, submittal letter, the applicant requested renewal of the operating license issued under Section 104b (Operating License No. DPR-16) of the Atomic Energy Act of 1954, as amended, for OCGS for a period of 20 years beyond the current license expiration date of midnight April 9, 2009. OCGS is located in Lacey Township, Ocean County, New Jersey, approximately two miles south of the community of Forked River, two miles inland from the shore of Barnegat Bay, and nine miles south of Toms River, New Jersey. The NRC issued the OCGS construction permit on December 15, 1964, and the OCGS operating license on July 2, 1991. OCGS is a single unit facility with a single-cycle, forced-circulation boiling water reactor (BWR)-2 and a Mark 1 containment. The nuclear steam supply system was furnished by General Electric (GE) and the balance of the plant was originally designed and constructed by Burns & Roe. OCGS's licensed power output is 1930 megawatt thermal with a gross electrical output of approximately 619 megawatt electric. The updated final safety analysis report (UFSAR) contains details concerning the plant and the site.

The license renewal process consists of two concurrent reviews: (1) a technical review of safety issues and (2) an environmental review. The NRC regulations found in 10 CFR Parts 54 and 51, respectively, set forth the requirements against which license renewal applications are reviewed. The safety review for the OCGS license renewal is based on the applicant's LRA and responses to the staff's requests for additional information. The applicant supplemented its LRA and provided clarifications through its responses to requests for additional information in audits, meetings, and docketed correspondence. Unless otherwise noted, the staff reviewed and considered information submitted through December 15, 2006, and after this date on a case-by-case basis depending on the stage of the safety review and on the volume and complexity of the information. The public may view the LRA and all pertinent information and materials, including the UFSAR, at the NRC Public Document Room on the first floor of One White Flint North, 11555 Rockville Pike, Rockville, MD 20852-2738 (301-415-4737 / 800-397-4209), and at the Lacey Branch - Ocean County Library, 10 East Lacey Road, Forked River, NJ 08731. In addition, the public may find the LRA, as well as materials related to the license renewal review, on the NRC Web Site at www.nrc.gov.

This SER summarizes the results of the staff's safety review of the LRA and describes the technical details considered in evaluating the safety aspects of the proposed operation for an additional 20 years beyond the term of the current operating license. The staff reviewed the LRA in accordance with NRC regulations and the guidance of NUREG-1800, Revision 1, "Standard Review Plan for Review of License Renewal Applications for Nuclear Power Plants" (SRP-LR), dated September 2005.

SER Sections 2 through 4 address the staff's review and evaluation of license renewal issues considered during the review of the application. Section 5 is reserved for the report of the Advisory Committee on Reactor Safeguards (ACRS). Conclusions of this report are presented in Section 6.

SER Appendix A contains a table that identifies the applicant's commitments for the renewal of the operating license. Appendix B provides a chronology of the principal correspondence between the staff and the applicant on the review of the application. Appendix C is a list of the principal contributors to this SER. Appendix D is a bibliography of the references in support of the review.

In accordance with 10 CFR Part 51, the staff prepared a draft, plant-specific supplement to NUREG-1437, "Generic Environmental Impact Statement for License Renewal of Nuclear Plants (GEIS)". This supplement discusses the environmental considerations for renewal of the OCGS license. The staff issued Draft Supplement 28 to NUREG-1437, "Generic Environmental Impact Statement for License Renewal of Nuclear Plants, Regarding Oyster Creek Generating Station, Draft Report for Comment," in June 2006.

1.2 License Renewal Background

Pursuant to the Atomic Energy Act of 1954, as amended, and NRC regulations, operating licenses for commercial power reactors are issued for 40 years. These licenses can be renewed for up to 20 additional years. The original 40-year license term was selected on the basis of economic and antitrust considerations rather than on technical limitations; however, some individual plant and equipment designs may have been engineered for an expected 40-year service life.

In 1982, the staff anticipated interest in license renewal and held a workshop on nuclear power plant aging. This workshop led the staff to establish a comprehensive program plan for nuclear plant aging research. With the results of that research, a technical review group concluded that many aging phenomena are readily manageable and pose no technical issues that would preclude life extension for nuclear power plants. In 1986, the staff published a request for comment on a policy statement that would address major policy, technical, and procedural issues related to license renewal for nuclear power plants.

In 1991, the staff published the license renewal rule in 10 CFR Part 54 (the Rule), (56 FR 64943 dated December 13, 1991). The staff participated in an industry-sponsored demonstration program to apply the Rule to a pilot plant and to gain experience necessary to develop implementation guidance. To establish a scope of review for license renewal, the Rule defined age-related degradation unique to license renewal; however, during the demonstration program, the staff found that adverse aging effects that occur to plant systems and components are managed during the period of initial license. In addition, the staff found that the scope of the review did not allow sufficient credit for existing programs, particularly the implementation of the Maintenance Rule, which also manages plant-aging phenomena. As a result, the staff amended the Rule in 1995 (60 FR 22461 dated May 8, 1995). The amended Rule established a regulatory process simpler, more stable, and more predictable than the previous Rule. In particular, the staff amended the Rule to focus on managing the adverse effects of aging rather than on identifying age-related degradation unique to license renewal. The staff initiated these Rule changes to ensure that important systems, structures, and components (SSCs) will continue to perform their intended functions during the period of extended operation. In addition, the revised Rule clarified and simplified the integrated plant assessment process to be consistent with the revised focus on passive, long-lived structures and components (SCs).

In parallel with these efforts, the staff pursued a separate rulemaking effort and developed an amendment to 10 CFR Part 51 to focus the scope of the review of environmental impacts of license renewal and fulfill the NRC's responsibilities under the National Environmental Policy Act of 1969.

1.2.1 Safety Review

License renewal requirements for power reactors are based on two key principles:

(1) The regulatory process is adequate to ensure that the licensing bases of all currently operating plants provide and maintain an acceptable level of safety, with the possible exception of the detrimental effects of aging on the functionality of certain SSCs, as well as a few other safety-related issues, during the period of extended operation.

(2) The plant-specific licensing basis must be maintained during the renewal term in the same manner and to the same extent as during the original licensing term.

In implementing these two principles, 10 CFR 54.4 defines the scope of license renewal as including those SSCs (1) that are safety-related, (2) whose failure could affect safety-related functions, and (3) that are relied on for compliance with NRC regulations for fire protection, environmental qualification (EQ), pressurized thermal shock (PTS), anticipated transient without scram (ATWS), and station blackout (SBO).

Pursuant to 10 CFR 54.21(a), an applicant for a renewed license must review all SSCs within the scope of the Rule to identify SCs subject to an aging management review (AMR). Those SCs subject to an AMR perform an intended function without moving parts or without a change in configuration or properties and are not subject to replacement based on a qualified life or specified time period. As required by 10 CFR 54.21(a), an applicant for a renewed license must demonstrate that the effects of aging will be managed in such a way that the intended function(s) of those SCs will be maintained, consistent with the current licensing basis (CLB), for the period of extended operation; however, active equipment is considered to be adequately monitored and maintained by existing programs. In other words, the detrimental effects of aging that may affect active equipment are more readily detectable and can be identified and corrected through routine surveillance, performance monitoring, and maintenance activities. The surveillance and maintenance activities programs for active equipment, as well as other aspects of maintaining the plant's design and licensing basis, are required throughout the period of extended operation.

Pursuant to 10 CFR 54.21(d), the LRA is required to include a UFSAR supplement with a summary description of the applicant's programs and activities for managing the effects of aging and an evaluation of time-limited aging analyses (TLAAs) for the period of extended operation.

License renewal also requires identification and updating of TLAAs. During the design phase for a plant, certain assumptions about the length of time that the plant can operate are incorporated into design calculations for several of the plant's SSCs. In accordance with 10 CFR 54.21(c)(1), the applicant must either show that these calculations will remain valid for the period of extended operation, project the analyses to the end of the period of extended operation, or demonstrate that the effects of aging on the intended function(s) will be adequately managed for the period of extended operation.

In 2001, the staff developed and issued Regulatory Guide 1.188, "Standard Format and Content for Applications to Renew Nuclear Power Plant Operating Licenses." This regulatory guide endorses Nuclear Energy Institute (NEI) 95-10, "Industry Guideline for Implementing the Requirements of 10 CFR Part 54 - The License Renewal Rule," dated March 2001. NEI 95-10 details an acceptable method of implementing the Rule. The staff also used the SRP-LR to review the application.

In the LRA, the applicant fully utilized the process defined in NUREG-1801, Revision 1, "Generic Aging Lessons Learned (GALL) Report," dated September 2005. The GALL Report provides the staff with a summary of staff-approved aging management programs (AMPs) for the aging of many SCs subject to an AMR. If an applicant commits to implementing these staff-approved AMPs, the time, effort, and resources used to review an applicant's LRA can be greatly reduced, thereby improving the efficiency and effectiveness of the license renewal review process. The GALL Report summarizes the aging management evaluations, programs, and activities credited for managing aging for most of the SCs used throughout the industry. The report also serves as a reference for both applicants and staff reviewers to quickly identify AMPs and activities that the staff determined can provide adequate aging management during periods of extended operation.

1.2.2 Environmental Review

Part 51 of 10 CFR governs environmental protection regulations. In December 1996, the staff revised the environmental protection regulations to facilitate the environmental review for license renewal. The staff prepared the GEIS to document its evaluation of the possible environmental impacts of renewed licenses for nuclear power plants. For certain types of environmental impacts, the GEIS establishes generic findings applicable to all nuclear power plants. These generic findings are codified in Appendix B to Subpart A of 10 CFR Part 51. Pursuant to 10 CFR 51.53(c)(3)(i), an applicant for license renewal may incorporate these generic findings in its environmental report. In accordance with 10 CFR 51.53(c)(3)(ii), an environmental report must also include analyses of environmental impacts that must be evaluated on a plant-specific basis (i.e., Category 2 issues).

In accordance with the National Environmental Policy Act of 1969 and the requirements of 10 CFR Part 51, the staff reviewed the plant-specific environmental impacts of license renewal, including whether the GEIS had not considered new and significant information. As part of its scoping process, the staff held a public meeting November 1, 2005, in Toms River, New Jersey, to identify environmental issues specific to the plant. The draft, plant-specific Supplement 28 to the GEIS, dated June 2006, documents the results of the environmental review and includes a preliminary recommendation on the license renewal action. The staff held another public meeting on July 12, 2006, in Toms River, New Jersey, to discuss draft GEIS Supplement 28. After considering comments on the draft, the staff published the final, plant-specific GEIS Supplement 28, on January 29, 2007.

1.3 Principal Review Matters

Part 54 of 10 CFR describes the requirements for renewing operating licenses for nuclear power plants. The staff performed its technical review of the LRA in accordance with NRC guidance and the requirements of 10 CFR Part 54. Section 54.29 of 10 CFR sets forth the standards for renewing a license. This SER describes the results of the staff's safety review.

Section 54.19(a) of 10 CFR requires license renewal applicants to submit general information. The applicant provided this general information in LRA Section 1. The staff reviewed LRA Section 1 and found that the applicant had submitted the information required by 10 CFR 54.19(a).

Section 54.19(b) of 10 CFR requires each LRA to include "conforming changes to the standard indemnity agreement, 10 CFR 140.92, Appendix B, to account for the expiration term of the proposed renewed license." In the LRA, the applicant stated the following regarding this issue:

> The current indemnity agreement (No. B-37) for Oyster Creek states in Article VII that the agreement shall terminate at the time of expiration of the licenses specified in Item 3 of the Attachment to the agreement. Item 3 of the Attachment to the indemnity agreement lists license number, DPR-16. Applicant requests that any necessary conforming changes be made to Article VII and Item 3 of the Attachment, and any other sections of the indemnity agreement as appropriate to ensure that the indemnity agreement continues to apply during both the terms of the current license and the terms of the renewed license. Applicant understands that no changes may be necessary for this purpose if the current license number is retained.

The staff intends to maintain the original license number upon issuance of the renewed license, if approved. Therefore, conforming changes to the indemnity agreement need not be made and the requirements of 10 CFR 54.19(b) have been met.

Section 54.21 of 10 CFR requires each LRA to contain (a) an integrated plant assessment, (b) a description of any CLB changes that occurred during the staff's review of the LRA, (c) an evaluation of TLAAs, and (d) a UFSAR supplement. LRA Sections 3, 4, and Appendix B address the license renewal requirements of 10 CFR 54.21(a) and (c). LRA Appendix A as supplemented by AmerGen letters 2130-06-20354 and 2130-06-20258 contains the license renewal requirements of 10 CFR 54.21(d).

Section 54.21(b) of 10 CFR requires that each year, following submission of the LRA, and at least three months before the scheduled completion of the staff's review, the applicant must submit an amendment to the LRA that identifies any changes to the facility's CLB materially affecting the contents of the LRA, including the UFSAR supplement. The applicant submitted an update to the LRA, by letter dated July 18, 2006, which summarizes the changes to the CLB that have occurred during the staff's review of the LRA. In a subsequent letter on December 3, 2006, as corrected by letter dated December 15, 2006, the applicant submitted an update to the LRA to incorporate changes from the October 2006 refueling outage. These submissions satisfy the requirements of 10 CFR 54.21(b).

Section 54.22 of 10 CFR 54.22 requires the LRA to include changes or additions to the technical specifications necessary to manage the effects of aging during the period of extended operation. In LRA Appendix D, the applicant stated that it had not identified any technical specification changes necessary to support issuance of the renewed operating license for OCGS. This statement adequately addresses the requirement specified in 10 CFR 54.22.

The staff evaluated the technical information required by 10 CFR 54.21 and 10 CFR 54.22 in accordance with NRC regulations and the guidance provided by the SRP-LR. SER Sections 2, 3, and 4 document the staff's evaluation of the technical information in the LRA.

As required by 10 CFR 54.25, the ACRS will issue a report to document its evaluation of the staff's review of the LRA and associated SER. SER Section 5 will incorporate the ACRS report, once it is issued. SER Section 6 documents the findings required by 10 CFR 54.29.

The final, plant-specific GEIS Supplement 28 will document the staff's evaluation of the environmental information required by 10 CFR 54.23 and will specify the considerations related to renewing the license for OCGS. The staff will prepare this supplement separately from this SER.

1.4 Interim Staff Guidance

The license renewal program is a living program. The staff, industry, and other interested stakeholders gain experience and develop lessons learned with each renewed license. The lessons learned address the staff's performance goals of safety and security; openness in the regulatory process; effectiveness, efficiency, realistic, and timely action; and excellence in agency management. Interim staff guidance (ISG) is documented for use by the staff, industry, and other interested stakeholders until it is incorporated into such license renewal guidance documents as the SRP-LR and the GALL Report.

The following table provides the current ISG, issued by the staff, as well as the SER sections in which the staff addresses each ISG issue.

ISG Issue (Approved ISG No.)	Purpose	SER Section
Nickel-alloy components in the reactor pressure boundary (LR-ISG-19B)	Cracking of nickel-alloy components in the reactor pressure boundary. ISG under development. NEI and EPRI-MRP will develop an augmented inspection program for GALL AMP XI.M11-B. This AMP will not be completed until the NRC approves an augmented inspection program for nickel-alloy base metal components and welds as proposed by EPRI-MRP.	N/A (PWRs only)
Corrosion of drywell shell in Mark I containments (LR-ISG-2006-01)	To address concerns related to corrosion of drywell shell in Mark I containments.	3.0.3.2.27 3.0.3.2.23 3.5 4.7.2

1.5 Summary of Open Items

As a result of its review of the LRA, including additional information submitted to the staff through July 10, 2006, the staff identified the following open items (OIs), which remained open when the SER with open items was issued in August 2006. An issue is considered open if the applicant has not presented sufficient information or if the staff has not completed its review. Each OI has been assigned a unique identifying number. By letters dated April 7, June 20, December 3, and December 15, 2006, and February 15, 2007, the applicant responded to those OIs. The staff reviewed these responses and closed each of the OIs. The basis for closing the OIs is as follows:

OI 4.7.2-1.1: (Section 4.7.2 - Drywell Corrosion)

In RAI 4.7.2-1 dated March 10, 2006, the staff requested that the applicant provide the following information: For the drywell corrosion during the late 1980s and the new corrosion found during the subsequent inspections, provide the process used to establish confidence that the sampling done to identify the areas of corrosion has been adequate.

In its response dated April 7, 2006, the applicant emphasized that it employs a robust process to establish confidence that the nature and locations of sampling done and areas considered for identifying the areas of corrosion have been adequate. The applicant stated that the elements of process had been developed over several years and defined in several technical documents submitted to the NRC in the 1990s. In addition, the applicant stated that OCGS has conducted extensive examinations to identify the cause of drywell corrosion, employed a robust sampling process, quantified with reasonable assurance the extent of drywell shell thinning due to corrosion, and assessed its impact on the drywell's structural integrity.

The staff's review of the applicant's response determined that there had been no UT measurements taken in the lower portion of the spherical area above the sand-pocket area. The staff requested that the applicant clarify its UT sampling plan for the entire drywell shell assessment.

In its supplemental response dated June 20, 2006, the applicant stated:

> A review of the drywell fabrication and installation details show that the welds that attach the 0.770 inches (the correct thickness is 0.770 inches, not 0.722 inch as indicated in the meeting notes) nominal plates to the 1.154 inch nominal plates at elevation 23 ft 6 7/8 inch are double bevel full penetration welds. The external edge of the 1.154 inches plates is tapered to 3 to 12 minimum as required by ASME Section VIII, Subsection UW-35, while the internal edge of the 1.154 inch plates are flush with the 0.770 inch plates. Thus there are no ledges that could retain water leakage and result in more severe corrosion than in areas included in the inspection program. Also, this joint is located below the equatorial center of the sphere. Therefore, in the event that water may run down the gap between the drywell shell and the concrete wall it would not collect on this joint.

> In 1991, Oyster Creek performed random inspections of the drywell shell. Ultrasonic testing inspections were conducted at 19 locations on either the 1.154 inch thick plates or on the 0.770 inch thick plates. The UT measurements were taken on a 6 inch x 6 inch grid (49 UTs) at each location. The UT measurement results show that thinning of the plates at these locations is less severe than the areas that are included in the corrosion-monitoring program. For this reason, the transition area was not added to the corrosion-monitoring program. Based on the above, AmerGen concludes that areas monitored under the drywell corrosion monitoring program bound the transition (from 1.154 inches to 0.770 inch thick plates) area of the drywell shell. Nevertheless, UT measurements will be taken on the 0.770 inch thick plate, just above the weld, prior to entering the period of extended operation.

> The measurements will be conducted at one location using the 6 inch x 6 inch grid. A second set of UT measurements will be taken two refueling outages later at the same location. The results of the measurements will be analyzed and evaluated to confirm that the rate of corrosion in the transition is bounded by the rate of corrosion of the monitored areas in the upper region of the drywell. If corrosion in the transition area is found to be greater than areas monitored in the upper region of the drywell, UT inspections in the transition area will be performed on the same frequency as those performed on the upper region of the drywell (every other refueling outage).

> Similarly, a review of fabrication and installation details of the containment drywell shell shows that the weld that connects the 2.625" knuckle plates to the 0.640"cylinder plates at elevation 71 ft 6 inch is a double bevel full penetration weld. The edges of the 2.625 inch plates were fabricated with a 3 to 12 taper to provide a smooth transition from the thicker to the thinner plate as required by ASME Section VIII, Subsection UE-35. Thus there are no ledges that could retain water leakage and result in more severe corrosion than the areas included in the inspection program.

In 1991, Oyster Creek performed random inspections of the drywell shell. Ultrasonic testing (UT) inspections were conducted at 18 locations on the 2.625 inch thick knuckle plate and at four (4) locations on the 0.640 inch thick cylinder plate. The UT measurements were taken on a 6 inch x 6 inch grid (49 UTs) at each location. The UT measurement results showed that thinning of the plates at these locations was less severe than the areas that are included in the corrosion monitoring program. For this reason the knuckle area was not added to the corrosion monitoring program. Based on the above, AmerGen concludes that areas monitored under the drywell corrosion monitoring program bound the knuckle area of the drywell shell. However, UT measurements will be taken above the 2.625 inch knuckle plate in the 0.640 inch thick plate prior to entering the period of extended operation.

The staff believes that random sampling of UT measurement is valuable if the likelihood of corrosion is almost equal at every place in the region considered for UT measurements. If the geometry of the region and water flow in the air gap suggest that one area is more likely to have corrosion than another then the sampling plan must consider areas more likely to have corrosion in addition to the randomly selected areas. If the water flow in the air gap is high, the applicant's argument that the weld transition will not allow water accumulation would be accurate. However, if the water flow is slow, the applicant's argument may not hold true. During the forthcoming outage, the applicant plans UT measurements at one location on each of the transition areas. The staff believes that measurement at four locations in each transition area would be more conservative. The locations along the thickness transition should be consistent with the areas that have large water accumulation and corrosion in the sand bed region. This item was identified as Open Item 4.7.2-1.1 in the SER with Open Items issued in August 2006.

The applicant updated the IWE Program Commitments in its December 3, 2006, submission (pages 73 and 74, items 10 and 11) with four separate sets of UT thickness measurements of the drywell shell at two areas of transition between shell plate thicknesses using a 6"x6" grid (i.e., four separate 49-point UT sets at the transition at elevation 23' 6 7/8" and four sets of UTs at elevation 71'-6"). The specific locations selected will be based on previous operational experience (i.e., biased toward areas that have experienced corrosion or exposure to water leakage). These measurements will be at the same locations prior to the period of extended operation and at the second refueling outage after the initial inspection. If corrosion in these transition areas is greater than in areas monitored in the upper drywell, UT inspections in the transition areas will be on the same frequency as those in the upper drywell (every other refueling outage). Of these four locations, there were UT measurements at two for each transition area during 2006 outage. These first-time readings show that the mean and individual thicknesses meet acceptance criteria with adequate margin. There will be UT measurements in the remaining two locations at each transition area during the next outage prior to the period of extended operation.

The staff finds that the applicant's actions to include in the program UT measurement of shell areas that may experience increased rates of corrosion resolves the staff concern. The basis for this finding is that the UT measurements should provide an adequate database to confirm whether the random sampling program for UT measurements is reasonably representative.

The staff, however, noted an inconsistency in the license renewal commitment list (pages 45 and 46, commitment number 27, "ASME Section XI, Subsection IWE," item numbers 10 and 11) where it states that the UT measurements will be at one location. In a letter dated December 15, 2006, the applicant noted the editorial error in its letter dated December 3, 2006. The applicant corrected the error by changing commitment 27 item numbers 10 and 11 from UT measurements at one location to UT measurements at four locations. Open Item OI 4.7.2-1.1 is closed.

In its letter dated February 15, 2007, the applicant revised a commitment (Commitment No. 27) by adding Item 21, which states that the performance of the full scope of drywell sand bed region inspections will be conducted every other refueling outage. The staff identified this commitment item as a license condition.

<u>OI 4.7.2-1.2:</u> (Section 4.7.2 - Drywell Corrosion)

In RAI 4.7.2-1 dated March 10, 2006, the staff requested that the applicant provide the following information: For the drywell corrosion during the late 1980s and the new corrosion found during the subsequent inspections, provide the process used to establish confidence that the sampling done to identify the areas of corrosion has been adequate.

The staff's review of the April 7, 2006, response determined that the most susceptible bays in the sand pocket region of the drywell shell had been incorporated in the sampling. However, it was not clear to the staff whether the junction at elevation 6' 10.25" had been represented in the sampling. To determine whether the readings are taken at the vulnerable locations and reliable techniques are used, the staff requested that the applicant explain why this area should not be included in the sampling plan.

In its response dated June 20, 2006, the applicant noted that the drywell construction and fabrication details show that the presence of the drywell skirt prevents moisture intrusion into the plate. The applicant also noted that AmerGen has extensively investigated drywell corrosion, including the embedded shell. Plant-specific and industry operating experience indicate that corrosion of the embedded steel in concrete is not significant because the shell is protected by the high alkalinity of concrete. Corrosion could become significant only if the concrete environment is aggressive. The applicant also stated that historical data show that the environment in the sand bed region is not aggressive, and thus any water in contact with the embedded shell is not aggressive. The data show that corrosion of the drywell shell in the sand bed region is galvanic and impurities like chlorides and sulfates are not fundamentally involved in the anodic and cathodic corrosion reactions. Thus, only limited corrosion is anticipated for the drywell embedded shell.

The applicant concluded that corrosion monitoring of the sand bed region of the drywell shell is bounding with respect to corrosion that may have occurred on the drywell embedded shell before 1992. After 1992, corrosion of the embedded shell has not been significant because of the mitigative measures implemented and the robust drywell corrosion AMP.

The staff understands the applicant's technical basis to support the applicant's view that the inaccessible portion of the drywell shell (i.e., embedded between the concrete floor inside, and concrete outside) is not likely to be subject to the same type of severe corrosion as experienced in the sand bed area. However, the general corrosion in the liner plates embedded in concrete of a number of pressurized water reactor (PWR) and BWR containments suggests that certain irregularities during the construction (i.e. foreign objects or voids in the concrete) could trigger corrosion not arrested by the concrete environment. This suggestion is particularly significant for the plates potentially subject to water seepage. The applicant's position that the uniformly reduced thickness used in the GE analysis compensates for any corrosion that may have occurred before the area was sealed in 1992 has some validity. This item was Identified as Open Item OI 4.7.2-1.2 in the SER with Open Items issued in August 2006.

During the October 2006 refueling outage, the applicant inspected the embedded drywell shell in the trenches in bays #5 and 17 after removing the filler material in the trenches. The applicant observed approximately 5 inches of standing water in the trench in bay #5, and the trench in bay #17 was damp. Applicant investigations concluded that the likely water sources were a deteriorated drainpipe connection and a void in the bottom of the Sub-Pile Room drainage trough or condensation within the drywell that either fell or washed down the inside of the drywell shell to the concrete floor. Water samples from the trench in bay #5 were tested and determined to be non-aggressive in pH (8.4 – 10.21), chlorides (13.6 – 14.6 ppm), and sulfates (228 – 230 ppm).

The applicant entered the condition into the corrective action process. Several corrective actions included repair of the trough concrete in the area under the reactor vessel to prevent water from migrating through the concrete and reaching the drywell shell and caulking of the interface between the drywell shell and the drywell concrete floor/curb including the trench areas. The trench bay in bay #5 also was excavated to uncover an additional 6 inches of the internal drywell shell surface for inspection and UT thickness measurement. A total of 584 UT thickness measurements were taken with a 6"x6" template within the two trenches. Forty-two additional UT measurements were taken in the newly exposed area in bay #5.

Visual examination of the drywell shell within the two trenches detected minor surface rust with no recordable corrosion on the shell inner surface. The UT measurements indicated that the drywell shell in the trench areas had experienced a 0.038" reduction in average thickness since 1986. Amergen concluded that the wall thinning was a result of corrosion on the exterior surface of the drywell shell in the sand bed region between 1986 and 1992 when the sand was still in place and the corrosion was known.

An engineering evaluation to determine the impact of the as-found water on the continued integrity of the drywell concluded that the measured water chemistry values and the lack of any indications of rebar degradation or concrete surface spalling suggest that the protective passive film established during concrete installation at the embedded steel/concrete interface is still intact and that significant corrosion of the drywell shell is not expected as long as this benign environment is maintained. More specifically, this engineering evaluation indicates that no significant corrosion of the inner surface of the embedded drywell shell is anticipated for the following reasons:

- The water in contact with the drywell shell has been in contact with the adjacent concrete, which is alkaline, increases the pH of the water, and inhibits corrosion. This high-pH water contains levels of impurities significantly below the Electric Power Research Institute (EPRI) embedded steel guidelines action level recommendations.

- Any new water (*e.g.*, reactor coolant) entering the concrete-to-shell interface (now minimized by repairs) also increases pH by its migration through and contact with concrete, creating a non-aggressive, alkaline environment.

- Minimal corrosion of the wetted inner drywell steel surface in contact with concrete is expected only during outages because the drywell is inerted with nitrogen during operations. Even during outages, shell corrosion losses are expected to be insignificant as the exposure time to oxygen is very limited and the water pH is expected to be relatively high. Also repairs/modifications during the 2006 outage will further minimize exposure of the drywell shell to oxygen.

After the UT thickness measurements during the 2006 outage of the newly-exposed shell area in bay #5, which had not been examined since initial construction, a reduction of average shell thickness of 0.041" was observed. The applicant maintains that, although no continuing corrosion is expected, there is sufficient margin for both the 1.154" thick plate and the 0.676" thick plate even assuming the same reduction until the end of the period of extended operation.

The applicant also has enhanced the AMP to require periodic inspection of the drywell shell subject to concrete (with water) environments in the internal embedded shell area. After each inspection, UT thickness measurements will be evaluated and compared to previous UT thickness measurements. If results are unsatisfactory additional corrective actions, as necessary, will maintain drywell shell integrity throughout the period of extended operation.

To investigate the feasibility of state-of-the-art non-destructive examination techniques to determine the condition of the embedded region, the applicant contacted EPRI and other utility owners that use these techniques. After discussions and findings, the applicant understood that a "guided wave" technology may be able to provide some qualitative information on whether the embedded shell has undergone corrosion; however, neither this nor any other known non-destructive methods could determine the thickness of the embedded drywell shell or the specific extent of corrosion.

Based on review of the applicant's evaluation of the condition of the inaccessible portion of drywell shell embedded in concrete, the applicant's actions to date, and the enhanced inspection program including a detailed UT measurement plan to which the applicant committed, the staff concludes with reasonable assurance that the environment in the region is sufficiently non-aggressive for no significant progressive corrosion. Therefore, the staff concern is resolved and Open Item 4.7.2-1.2 is closed.

In its letter dated February 15, 2007, the applicant change a commitment (Commitment No. 27) by adding Item 20, which states AmerGen is committed to perform visual and UT inspections of the drywell shell in the inspection trenches in drywell bays #5 and #17. AmerGen will monitor the two trenches for the presence of water during each refueling outage. The staff identified this commitment item as a license condition.

OI 4.7.2-1.3: (Section 4.7.2 - Drywell Corrosion)

In RAI 4.7.2-1 dated March 10, 2006, the staff requested that the applicant provide the following information: A summary of the factors considered in establishing the minimum required drywell thickness.

In its response dated April 7, 2006, the applicant explained that the factors considered in establishing the minimum required drywell thickness at various elevations of the drywell are described in detail in engineering analyses documented in two GE reports, Index Nos. 9-1, 9-2, and 9-3, 9-4.

In the applicant's discussion, a summary of the methods and assumptions used in the buckling analysis of the shell in the sand-pocket area has been given. Although the NRC has not approved ASME Code Case N-284 for use on a generic basis, the staff does not take exception to the use of average compressive stress across the metal thickness for buckling analysis of the as-built shell. However, if the corrosion has reduced the strength of the remaining metal through the cross section, this use may not be valid. The staff requested that the applicant address this issue.

In its response dated June 20, 2006, the applicant discussed its use of ASME Code Case N-284:

> Although Revision 1 of Code Case 284 had not yet been issued when the report (An ASME Section VIII Evaluation of Oyster Creek Drywell for Without Sand Case, Part II - Stability Analysis," GE Report, Index No. 9-4, Revision 0, DRF # 00664) was written, the authors consulted with the primary author of the revision. Based on those discussion, the plasticity correction factors used in the evaluation are the same as those in Figure 1610-1 of Code Case N-284 Revision 1.

The applicant stated that the technical approach used in the stability evaluation of Reference 2 is entirely consistent with the guidelines in ASME Code Case N-284, Revision 1. In addition, the applicant concluded that the corrosion on the outside surface of the shell will not introduce eccentricities that would significantly impact the "e/t" value of 1.0 assumed in ASME Code Case N-284. The applicant also stated that it expected additional eccentricity from shell corrosion in service to be accommodated within the allowable limit for imperfections.

The staff believed that the applicant provided a thorough explanation of the factors considered in applying the ASME Code Case N-284-1 for buckling analysis of the corroded shell in the sand bed area of the drywell shell. However, the applicant did not address whether it is appropriate to assume the same strength across the corroded section of the shell. The incorporation of the "e/t" corrosion concept with a representative distribution of strength along the corroded section that recognize the lower strength at the corroded side and full strength at the inside surface, could support the claim of conservatism in the analysis. This item was identified as Open Item OI 4.7.2-1.3 in the SER with Open Items issued in August 2006.

On further evaluation of the applicant's information, the staff concludes that the stability evaluation was consistent with the guidelines of ASME Code Case N-284-1. The staff's concern about use of the same section strength across the corroded section of the shell is addressed by Code Case N-284-1, which uses conservative assumptions to determine shell capacity reduction factors (*i.e.*, assumption of imperfection limit indicated by parameter "e/t" to be 1.0 in the code case) expected to compensate reasonably for such use of the same section strength. In addition, the applicant conservatively assumed the local corroded thickness for the entire drywell shell region and demonstrated that the code-allowable stresses were satisfied consistently with the guidelines of the code case. Thus, this analysis adds a margin of safety for the drywell stability evaluation. On this basis, the staff believes that the stability evaluation method is adequate and acceptable, and the staff's concern is resolved. Open Item 4.7.2-1.3 is closed.

OI 4.7.2-1.4: (Section 4.7.2 - Drywell Corrosion)

In RAI 4.7.2-1 dated March 10, 2006, the staff requested that the applicant provide the following information: A summary of the factors considered in establishing the minimum required drywell thickness.

In its response dated April 7, 2006, the applicant explained that the factors considered in establishing the minimum required drywell thickness at various elevations of the drywell are described in detail in engineering analyses documented in two GE reports, Index Nos. 9-1, 9-2, and 9-3, 9-4.

For the localized thin areas, the applicant uses the provision of NE-3213.10 of Subsection NE of ASME Code Section III. This provision, although not directly applicable to the randomly thin areas caused by corrosion, if used with care and adequate conservatism, could provide information about the primary stress levels at the junction of the thin and thick areas. The staff requested that the applicant provide a summary of the process used to address this issue.

In its response dated June 20, 2006, the applicant noted that "although provisions in ASME Code Section III, Subsection NE-3213.10 are not directly applicable to the randomly thin areas caused by corrosion, AmerGen believes that the provisions are applicable to the analysis of Oyster Creek drywell shell based on the following:

- The stress analysis of Oyster Creek drywell presented in Reference 1 satisfies the local primary stress requirements of NE-3213.10. Conservatism in the allowable primary stress intensity value, the assumed peak pressure during the LOCA condition and the assumption of local corroded thickness in the entire region of the drywell provide additional structural margin.

- The Code primary stress limits are satisfied in the corroded condition and the number of fatigue cycles is small, the surface discontinuities from corrosion do not represent a significant structural integrity concern.

- The applicant indicated that UT measurements of the drywell shell above the sand bed region had shown that the measured general thickness contains significant margin. The applicant stated that the ongoing corrosion in that region is insignificant and that the margin could be applied to offset uncertainties related to surface roughness.

- The applicant stated that UT measurements of the drywell shell in the sand bed region show that the measured general thickness is greater than the 0.736'" thickness assumed in the buckling analysis by significant margins except in two bays, 17 and 19. (Refer to response to RAI 4.7.2-1(d), Table-2). The margin in the general thickness of the two bays is 0.074" and 0.064" respectively. As significant additional corrosion is not expected in the sand bed region, the applicant applied the margin to offset uncertainties related to the surface roughness.

Because the staff had not completed its evaluation, this item was identified as Open Item OI 4.7.2-1.2 in the SER with Open Items issued in August 2006.

After further evaluation of the applicant's justification, the staff accepts the use of the NE-3213.10 provisions of Subsection NE of ASME Code Section III. The staff acceptance is based on the applicant's conservative approaches to its determination of the allowable shell capacity. Specifically, the applicant demonstrated acceptable shell capacity based on a conservative LOCA peak internal pressure (i.e., peak internal pressure of 62 psi in the evaluation versus the 44 psi peak internal pressure in an Oyster Creek specific calculation approved by the NRC in 1993), use of a local corroded thickness for the entire region of the drywell, and compliance with local primary stress code limits in the corroded condition. In addition, the applicant expects its enhanced actions to prevent significant additional corrosion in the sand bed region. With this information, the staff's concern is resolved and Open Item 4.7.2-1.4 is closed.

OI 4.7.2-3: (Section 4.7.2 - Drywell Corrosion)

In RAI 4.7.2-3 dated March 10, 2006, the staff noted that leakage from the refueling seal has been identified as one of the reasons for accumulation of water and contamination of the sand-pocket area. The refueling water passes through the gap between the shield concrete and the drywell shell in the long length of inaccessible areas. As there is a potential for corrosion, ASME Code Subsection IWE would require augmented inspection of this area. The staff requested that the applicant provide a summary of inspections (visual and NDE) and mitigating actions to prevent water leaks from the refueling seal components.

In its response dated April 7, 2006, the applicant stated that the refueling seals at OCGS consist of stainless steel bellows. In the mid-to-late 1980s, GPU conducted extensive visual and NDE inspections to determine the source of water intrusion into the seismic gap between the drywell concrete shield wall and the drywell shell and accumulation in the sand bed region. The inspections concluded that the refueling bellows (seals) were not the source of water leakage. The bellows were repeatedly tested by helium (external) and air (internal) with no indication of leakage. Furthermore, any minor leakage from the refueling bellows would be collected in a concrete trough below the bellows. The concrete trough is equipped with a drain line that would direct any leakage to the reactor building equipment drain tank and prevent it from entering the seismic gap. The drain line has been checked before refueling outages to confirm that it is not blocked. The only other seal is the gasket for the reactor cavity steel trough drain line. This gasket was replaced after the tests showed that it was leaking. However, the gasket leak was ruled out as the primary source of water observed in the sand bed drains because there is no clear leakage path to the seismic gap. Minor gasket leaks would be collected in the concrete trough below the gasket and would be removed by the drain line like leaks from the refueling bellows.

In addition, the applicant noted that additional visual and NDE (dye penetrant) inspections on the reactor cavity stainless steel liner had identified a significant number of cracks, some throughwall. Engineering analysis concluded that the cracks were most probably caused by mechanical impact or thermal fatigue, not intergranular stress corrosion cracking (IGSCC). These cracks were determined to be the source of refueling water that passed through the seismic gap. To prevent leakage through the cracks, GPU installed an adhesive-type stainless steel tape to bridge any observed large cracks and subsequently applied a strippable coating. This repair greatly reduced leakage and was implemented every refueling outage while the reactor cavity was flooded.

The applicant noted that OCGS has a long-time commitment to monitor the sand bed region drains for water leakage. A review of plant documentation provided no objective evidence that the commitment had been implemented since 1998. OCGS Issue Report No. 348545 was issued, in accordance with the corrective action process, to document the lapse in implementing the commitment and to reinforce strict compliance with commitment implementation in the future, including during the period of extended operation.

The applicant also committed (Commitment No. 27) to augmented inspections of the drywell in accordance with ASME Code Section XI, Subsection IWE. These inspections consist of UT examinations of the upper region of the drywell and visual examinations of the protective coating on the exterior of the drywell shell in the sand bed region. UT measurements will supplement the visual inspection of the coating measurements from inside the drywell once before entering the period of extended operation and every 10 years during the period of extended operation.

The staff's review of the applicant's response determined that the epoxy coating applied in the sand-bed region of the shell has a limited life and that water leakage from the air gap has not been prevented. With these observations, the staff requested that the applicant provide a systematic program of examination of the coating for confidence that the preventive measure is adequately implemented at all locations in the sand-pocket areas.

In its response dated June 20, 2006, the applicant committed to monitoring the sand bed region drains on a daily basis during refueling outages and take the following actions if water is detected. The following actions will be completed prior to exiting the outage:

- The source of water will be investigated and diverted, if possible, from entering the gap between the drywell shell and the drywell shield wall.

- The water will be chemically analyzed to aid in determining the source of leakage.

- A remote inspection will be performed in the trough drain area to determine if the trough drains are operating properly.

- The condition of the coating and the moisture barrier (seal) in the affected bays will be inspected.

- If the coating is degraded and visual inspection indicates corrosion is taking place, then UT thickness measurements will be taken in the affected areas of the sand bed region. The measurements will be taken from either inside or outside the drywell to ensure that the shell thickness in areas affected by water leakage is measured. UT thickness measurements and evaluation will be consistent with the existing program.

- The degraded coating and/or the seal will be repaired in accordance with station procedures.

- UT measurements will be taken in the upper region of the drywell consistent with the existing program.

The applicant also committed (Commitment No. 27) to monitor the sand bed region drains quarterly during the operating cycle. The applicant stated that, if water is detected, actions listed below will be taken. Actions that can only be completed during an outage will be completed during the next scheduled refueling outage.

- The leakage rate will be quantified to determine a representative flow rate. The leakage rate will be trended.

- The source of water will be investigated and diverted, if possible, from entering the gap between the drywell shell and the drywell shield wall.

- The water will be chemically analyzed to determine the source of leakage.

- The condition of the coating and the moisture barrier (seal) in the affected bays will be inspected during the next refueling outage or an outage of opportunity.

- If the coating is degraded and visual inspection indicates corrosion has taken place, then UT thickness measurements will be taken in the affected areas of the sand bed region from either inside or outside the drywell to ensure that the shell thickness in areas affected by water leakage is measured. UT thickness measurements and evaluation of the results will be consistent with the existing program.

- UT measurements will be taken in the upper region of the drywell consistent with the existing program.

- The degraded coating or the seal will be repaired in accordance with station procedures.

The staff believes that applicant had not provided sufficient information regarding the extent that coated surfaces will be examined during each inspection. This item was identified as Open Item OI 4.7.2-3 in the SER with Open Items issued in August 2006.

In a letter dated June 23, 2006, the applicant committed to monitoring of the coating on the drywell shell exterior in the sand bed region as part of its ASME Section XI, Subsection IWE Program and of its Protective Coating Monitoring and Maintenance Program. The applicant committed to additional visual inspections of the epoxy coating in all 10 drywell bays at least once prior to the period of extended operation. In a letter dated December 3, 2006, the applicant stated that 100 percent of the epoxy coating had been inspected during the October 2006 outage with no evidence of flaking, blistering, peeling, discoloration, or other signs of coating distress. The staff finds that these commitments with the IWE program and the absence of evidence of coating deterioration in the October 2006 inspection resolve the concern over the extent of coatings inspections. The staff's concern is resolved and Open Item 4.7.2-3 is closed.

1.6 Summary of Confirmatory Items

The staff's review of the LRA, including additional information submitted to the staff through December 15, 2006, identified no confirmatory items (CIs). An issue was considered confirmatory if the staff and the applicant have reached a satisfactory resolution, but such information has not yet been submitted to the staff.

1.7 Summary of Proposed License Conditions

As a result of its review of the LRA, recommendations from the Advisory Committee on Reactor Safeguards, and subsequent information and clarifications from the applicant, the staff, at present, proposes seven license conditions.

The first license condition requires the applicant to include the UFSAR supplement required by 10 CFR 54.21(d) in the next UFSAR update, as required by 10 CFR 50.71(e), following the issuance of the renewed license.

The second license condition requires future activities identified in the UFSAR supplement to be completed prior to entering and during the period of extended operation.

The third license condition requires all surveillance capsules placed in storage to be maintained for future insertion. Any changes to storage requirements must be approved by the staff as required by 10 CFR Part 50, Appendix H.

The fourth license condition requires the applicant to perform full scope inspections of the drywell sand bed region every other refueling outage.

The fifth license condition requires the applicant to monitor drywell trenches every refueling outage to identify and eliminate the sources of water and receive NRC approval prior to restoring the trenches to their original design configuration.

The sixth license condition requires the applicant to perform an engineering study prior to the period of extended operation to identify options to eliminate or reduce the leakage in the OCGS refueling cavity liner.

The seventh license condition requires the applicant to perform a 3-D (dimensional) finite-element analysis of the drywell shell prior to entering the period of extended operation.

SECTION 2

STRUCTURES AND COMPONENTS SUBJECT TO AGING MANAGEMENT REVIEW

2.1 Scoping and Screening Methodology

2.1.1 Introduction

Title 10, Section 54.21 of the *Code of Federal Regulations* (10 CFR Part 54.21), "Contents of Application Technical Information," requires each license renewal application (LRA) to contain an integrated plant assessment (IPA) listing those structures and components (SCs) subject to an aging management review (AMR) from all of the systems, structures, and components (SSCs) within the scope of license renewal in accordance with 10 CFR 54.4.

In LRA Section 2.1, "Scoping and Screening Methodology," the applicant described the methodology used to identify the SSCs at the Oyster Creek Generating Station (OCGS) within the scope of license renewal and the SCs subject to an AMR. The staff reviewed the AmerGen Energy Company, LLC (AmerGen or the applicant) scoping and screening methodology to determine whether it meets the scoping requirements of 10 CFR 54.4(a) and the screening requirements of 10 CFR 54.21.

In developing the scoping and screening methodology for the LRA, the applicant considered the requirements of 10 CFR 54, "Requirements for Renewal of Operating Licenses for Nuclear Power Plants," (the Rule), statements of consideration related to the Rule, and the guidance of Nuclear Energy Institute (NEI) 95-10, "Industry Guideline for Implementing the Requirements of 10 CFR Part 54 - The License Renewal Rule," Revision 5. Additionally, in developing this methodology, the applicant considered the correspondence between the staff and other applicants and/or the NEI.

2.1.2 Summary of Technical Information in the Application

LRA Sections 2.0 and 3.0 provide the technical information required by 10 CFR 54.21(a). LRA Section 2.1 describes the process to identify SSCs meeting the license renewal scoping criteria under 10 CFR 54.4(a) and the process to identify SCs subject to an AMR, as required by 10 CFR 54.21(a)(1). In addition, the applicant provided the results of the process to identify the SCs subject to an AMR in the following LRA sections:

- Section 2.2, "Plant Level Scoping Results"
- Section 2.3, "Scoping and Screening Results: Mechanical"
- Section 2.4, "Scoping and Screening Results: Structures"
- Section 2.5, "Scoping and Screening Results: Electrical Components"

LRA Section 3, "Aging Management Review Results," contains the applicant's aging management results in the following LRA sections:

- Section 3.1, "Aging Management of Reactor Vessel, Internals, and Reactor Coolant Systems"
- Section 3.2, "Aging Management of Engineered Safety Features Systems"
- Section 3.3, "Aging Management of Auxiliary Systems"
- Section 3.4, "Aging Management of Steam and Power Conversion System"
- Section 3.5, "Aging Management of Containment, Structures, Component Supports, and Piping and Component Insulation"
- Section 3.6, "Aging Management of Electrical Components"

LRA Section 4.0, "Time-Limited Aging Analyses," contains the applicant's identification and evaluation of time-limited aging analyses (TLAAs).

2.1.3 Scoping and Screening Program Review

The staff evaluated the LRA scoping and screening methodology in accordance with the guidance of Section 2.1, "Scoping and Screening Methodology," of NUREG-1800, Revision 1, "Standard Review Plan for Review of License Renewal Applications for Nuclear Power Plants," (SRP-LR). The following regulations form the basis for the acceptance criteria for the scoping and screening methodology review:

- 10 CFR 54.4(a), as it relates to the identification of plant SSCs within the scope of the Rule
- 10 CFR 54.4(b), as it relates to the identification of the intended functions of plant structures and systems within the scope of the Rule
- 10 CFR 54.21(a)(1) and (a)(2), as they relate to the methods utilized by the applicant to identify plant SCs subject to an AMR

As parts of the applicant's scoping and screening methodology, the staff reviewed the activities described in the following sections of the LRA using the guidance of SRP-LR:

- Section 2.1 to ensure that the applicant described a process for identifying SSCs within the scope of license renewal in accordance with the requirements of 10 CFR 54.4(a).
- Section 2.2 to ensure that the applicant described a process for determining SCs subject to an AMR in accordance with the requirements of 10 CFR 54.21(a)(1) and (a)(2).

In addition, the staff conducted a scoping and screening methodology audit at OCGS in New Jersey during the week of September 19 through 23, 2005. The audit focused on ensuring that the applicant had developed and implemented adequate guidance to conduct the scoping and screening of SSCs in accordance with the methodologies described in the LRA and the requirements of the Rule. The staff reviewed implementation of the project level instructions and position papers describing the applicant's scoping and screening methodology. In addition, the staff conducted detailed discussions with the applicant on the implementation and control of the license renewal programs and reviewed administrative control documentation and selected design documentation used by the applicant during the scoping and screening process. The staff reviewed the applicant's processes for quality assurance (QA) as to development of the LRA. The staff evaluated the quality attributes of the applicant's aging management program (AMP) activities described in LRA Appendix B, "Aging Management Programs." The staff also reviewed the training and qualification of the LRA development team. The staff reviewed scoping and screening results reports for the isolation condenser system (ICS) and reactor building to ensure that the applicant had appropriately implemented the methodology outlined in the administrative controls and that the results were consistent with the current licensing basis (CLB) documentation. The staff documented its review in an audit trip report issued on October 21, 2005. The report identified several issues which required additional information from the applicant prior to completion of the review.

2.1.3.1 Implementation Procedures and Documentation Sources Used for Scoping and Screening

The staff reviewed the applicant's scoping and screening implementation procedures to verify that the process used to identify SCs subject to an AMR was consistent with the LRA and the SRP-LR. Additionally, the staff reviewed the scope of CLB documentation sources and the process used by the applicant to ensure that CLB commitments had been appropriately considered and that the applicant had adequately implemented the procedural guidance during the scoping and screening process.

2.1.3.1.1 Summary of Technical Information in the Application

In LRA Section 2.1.2, "Information Sources Used for Scoping and Screening," the applicant reviewed the following information sources during the license renewal scoping and screening process:

- design basis documents (DBDs)
- component record list (CRL)
- updated final safety analysis report (UFSAR)
- fire hazards analysis report
- engineering drawings, evaluations, and calculations
- environmental qualification master list
- maintenance rule database
- NRC safety evaluation reports

The license renewal boundary drawings (LRBDs) show the systems within the scope of license renewal highlighted in color.

2.1.3.1.2 Staff Evaluation

Scoping and Screening Implementation Procedures. The staff reviewed the following scoping and screening methodology implementation procedures:

- Position Paper (PP)-01, "License Renewal Systems & Structures," Revision 3
- PP-02, "10 CFR 54.4(a)(1) Safety Related Systems and Structures," Revision 2

- PP-03, "10 CFR 54.4(a)(2) Systems and Structures," Revision 3
- PP-04, "Systems and Structures Relied Upon to Demonstrate Compliance with 10 CFR 50.63 - Station Blackout," Revision 2
- PP-05, "Systems and Structures Relied Upon to Demonstrate Compliance with 10 CFR 50.62 - ATWS," Revision 1
- PP-06, "Systems and Structures Relied Upon to Demonstrate Compliance with 10 CFR 50.49 - Environmental Qualification," Revision 1
- PP-07, "Systems and Structures Relied Upon to Demonstrate Compliance with 10 CFR 50.48 - Fire Protection," Revision 3
- PP-08, "Structures, Components and Commodity Types with Active, Passive, Short Lived Determinations and Intended Functions," Revision 2
- PP-13, "Abnormal Operating Occurrence," Revision 2
- Project Level Instruction (PLI)-02, "Scoping of Systems and Structures," Revision 4
- PLI-03, "Screening of Systems, Structures and Components," Revision 2
- PLI-04, "Boundary Drawings," Revision 2

The staff found the overall process to implement 10 CFR 54 requirements included in the PLIs. Guidance for determining plant SSCs within the scope of the Rule, including guidelines for determining which component types of the SCs within the scope of license renewal were subject to an AMR, were found by the staff in the PPs. During the review of these procedures, the staff focused on the consistency of the detailed procedural guidance with information in the LRA, including in the implementation of NRC staff positions documented in the SRP-LR and interim staff guidance (ISG) documents.

After reviewing the LRA and supporting documentation, the staff finds the scoping and screening methodology instructions consistent with LRA Section 2.1. The applicant's methodology has sufficient detail for concise guidance on the scoping and screening implementation process followed during LRA activities.

Sources of Current Licensing Basis Information. The staff reviewed the scope and depth of the applicant's CLB information to verify that the applicant's methodology had comprehensively identified SSCs within the scope of license renewal as well as components types requiring an AMR. As defined in 10 CFR 54.3(a), the CLB is applicable NRC requirements, written licensee commitments for ensuring compliance with, and operation within, applicable NRC requirements, and plant-specific design bases docketed and in effect. The CLB includes certain NRC regulations, orders, license conditions, exemptions, technical specifications, design-basis information documented in the most recent UFSAR, and licensee commitments made in such docketed licensing correspondence as licensee responses to NRC bulletins, generic letters, and enforcement actions as well as licensee commitments documented in NRC safety evaluations or licensee event reports.

During the audit, the staff reviewed pertinent information sources utilized by the applicant. The staff reviewed samples of information utilized by the applicant, including the

UFSAR, DBDs, controlled plant reference drawings, LRBDs, and Maintenance Rule information. In addition, the applicant developed and implemented a CLB database comprised of primarily licensing correspondence, UFSAR, technical specifications, fire hazards analysis, safety evaluations, and design documentation. This database enabled the applicant to search specific keywords and phrases to find licensing references applicable to license renewal. The applicant formally trained the license renewal staff on the CLB database and described the contents and practical experience in its use. Training lesson plans reviewed by the staff during the audit contained detailed information on important definitions related to the licensing basis, descriptions of the sources of documents which comprised the CLB, and descriptions of the programs and processes that contain the CLB source information. The applicant's detailed PLI-02 Section 6.0 requires use of the CLB source information in developing scoping evaluations. The applicant used the CLB electronic database, in part, for this process requirement.

The CRL is the applicant's primary repository for component safety classification information. During the audit, the staff reviewed the applicant's administrative controls for CRL safety classification data and has determined that the applicant had established adequate measures to control data integrity and reliability. Therefore, the staff concludes that the CRL provided a sufficiently controlled source of component data to support scoping and screening evaluations.

During the staff's review of the applicant's CLB evaluation process, the applicant discussed updates to the CLB and the process for their adequate incorporation into the license renewal process. The applicant provided the staff with PLI-16 and discussed the process defined for such updates. As part of the license renewal effort, the applicant ensured that all engineering change requests approved up to within three months of the LRA submission that could have affected it had been factored in. In addition, PLI-16 guides the evaluation of CLB change documentation that could impact the LRA, describes the process for annual updates to the LRA, and includes a series of checklists to facilitate the evaluation and ensure adequate documentation of the results.

The staff concludes that LRA Section 2.1 provides a description of the CLB and related documents used during the scoping and screening process consistent with SRP-LR guidance. In addition, the staff reviewed technical reports supporting identification of SSCs relied upon for compliance with the safety-related criteria, nonsafety-related criteria, and the five regulated events of 10 CFR 54.4(a). PLI-02 and PLI-16 comprehensively lists documents supporting scoping and screening evaluations. The staff finds these design documentation sources useful in ensuring that the initial scope of SSCs identified by the applicant is consistent with the plant's CLB.

2.1.3.1.3 Conclusion

On the basis of review of information in LRA Section 2.1, the detailed scoping and screening implementation procedures, and the results from the scoping and screening audit, the staff concludes that the applicant's scoping and screening methodology had considered CLB information consistently with SRP-LR and NEI 95-10 guidance and is, therefore, acceptable.

2.1.3.2 Quality Controls Applied to LRA Development

2.1.3.2.1 Staff Evaluation

The staff reviewed the applicant's quality controls to ensure that scoping and screening methodologies in the LRA had been adequately implemented. Although the applicant did not develop the LRA under a 10 CFR 50, Appendix B, QA program, the applicant utilized the following QA processes during the LRA development:

- The scoping and screening methodology was governed by written procedures, guidelines, PPs, PLIs, and project checklist packages.
- The applicant studied staff requests for additional information (RAIs) from the Dresden, Quad Cities, Nine Mile Point, and Beaver Valley plants to ensure that applicable issues were addressed in the OCGS LRA.
- The LRA was examined and approved by the applicant's Nuclear Safety Review Board and Plant Operations Review Committee.
- The applicant planned to retain certain license renewal documents as quality records or control documents.
- The applicant performed six independent party examinations of LRA development activities.
- Nuclear Oversight performed two self-assessments of the implementation of LRA.

2.1.3.2.2 Conclusion

On the basis of review of pertinent LRA development guidance, discussion with the applicant's license renewal personnel, and review of the quality audit reports, the staff concludes that these QA activities provided additional assurance that LRA development activities had been in accordance with the LRA descriptions.

2.1.3.3 Training

2.1.3.3.1 Staff Evaluation

The staff reviewed the applicant's training process for consistent and appropriate performance of the guidelines and methodology for scoping and screening. PLI-12 guided the training of the applicant's license renewal project team and site personnel and required them to review applicable license renewal regulations, NEI 95-10, and associated procedures. The applicant developed periodic production meetings in which the license renewal project team members shared their knowledge and experience. The staff reviewed the training records of the applicant's license renewal personnel and noted no discrepancies.

2.1.3.3.2 Conclusion

Based on discussions with the applicant's license renewal personnel responsible for the scoping and screening process and a review of selected documentation supporting the process, the staff concludes that the applicant's personnel understood the requirements and adequately implemented the scoping and screening methodology documented in the LRA. The staff concludes that the license renewal personnel were adequately trained and qualified for license renewal activities.

2.1.3.4 Conclusion of Scoping and Screening Program Review

On the basis of review of information in LRA Section 2.1, review of the applicant's detailed scoping and screening implementation procedures, discussions with the applicant's LRA personnel, and review of the results from the scoping and screening audit, the staff concludes that the applicant's scoping and screening program is consistent with SRP-LR guidance and, therefore, acceptable.

2.1.4 Plant Systems, Structures, and Components Scoping Methodology

In LRA Section 2.1, the applicant described the methodology for scoping SSCs pursuant to 10 CFR 54.4(a) and the scoping process for the plant in terms of systems and structures, identified system/structure level functions, and evaluated these functions against the 10 CFR 54.4(a) scoping criteria to determine whether they perform a license renewal intended function. The applicant evaluated the components in the systems and structures within the scope of license renewal. The in-scope boundary was depicted on the LRBDs. The applicant's scoping methodology, as described in the LRA, is discussed in the sections below.

2.1.4.1 Application of the Scoping Criteria in 10 CFR 54.4(a)(1)

2.1.4.1.1 Summary of Technical Information in the Application

In LRA Section 2.1.5.1, "Safety Related - 10 CFR 54.4(a)(1)," the applicant described the 10 CFR 54 scoping methodology and the 10 CFR 54.4(a)(1) safety-related criteria. The applicant stated that safety-related SCs are identified in the CRL and that safety-related classifications for SSCs are based on descriptions and analyses in the UFSAR or on DBDs like engineering drawings, evaluations or calculations. SSCs identified as safety-related in the UFSAR, in DBDs, or in the CRL were classified under 10 CFR 54.4(a)(1) and included within the scope of license renewal. The applicant also confirmed that all plant conditions, including normal operation, abnormal operational transients, design-basis accidents, internal and external events, and natural phenomena for which the plant must be designed, had been considered for license renewal scoping under 10 CFR 54.4(a)(1) criteria.

The CLB definition of "safety-related" is not identical to the definition in the Rule. The applicant evaluated the differences between the CLB and Rule definitions and documented the evaluation in LRA Section 2.1.3.2, "Identification of Safety-Related Systems and Structures," as well as in PP-02 and PP-13.

2.1.4.1.2 Staff Evaluation

Under 10 CFR 54(a)(1), the applicant must consider all safety-related SSCs relied upon to remain functional during and following a design basis event (DBE) to ensure (1) the integrity of the reactor coolant pressure boundary, (2) the ability to shut down the reactor and maintain it in a safe shutdown condition, or (3) the capability to prevent or mitigate the consequences of accidents that could result in potential offsite exposures comparable to those in 10 CFR 50.34(a)(1), 10 CFR 50.67(b)(2), or 10 CFR 100.11.

As to identification of DBEs, SRP-LR Section 2.1.3 states:

> The set of DBEs as defined in the Rule is not limited to Chapter 15 (or equivalent) of the UFSAR. Examples of DBEs that may not be described in this chapter include external events, such as floods, storms, earthquakes, tornadoes, or hurricanes, and internal events, such as a high energy line break. Information regarding DBEs as defined in 10 CFR 50.49(b)(1) may be found in any chapter of the facility UFSAR, the Commission's regulations, NRC orders, exemptions, or license conditions within the CLB. These sources should also be reviewed to identify SSCs relied upon to remain functional during and following DBEs (as defined in 10 CFR 50.49(b)(1)) to ensure the functions described in 10 CFR 54.4(a)(1).

The applicant scoped SSCs for the 10 CFR 54.4(a)(1) criterion following PP-01, -02, -13, and PLI-02, which guided the preparation, review, verification, and approval of the scoping evaluations to ensure adequate results. The staff reviewed these guidance documents governing the applicant's evaluation of safety-related SSCs and sampled the applicant's scoping results reports to ensure that the methodology had been implemented in accordance with those written instructions. In addition, the staff discussed the methodology and results with the applicant's personnel responsible for the evaluations.

Specifically, the staff reviewed a sample of the license renewal scoping results for the ICS and the reactor building for additional assurance that the applicant has adequately implemented their safety-related scoping methodology. The staff verified that the scoping results for the sampled system and structure had been developed consistently with the methodology, that the SSCs credited for performing intended functions had been identified, and that the bases for the results as well as the intended functions had been adequately described. The staff verified that the applicant had identified and used pertinent engineering and licensing information to identify the SSCs required to be within the scope of license renewal in accordance with 10 CFR 54.4(a)(1).

To help facilitate the identification of SSCs within the scope of license renewal in accordance with 10 CFR 54.4(a), the applicant developed a license renewal database with detailed design description information about each plant system and structure and their relevant functions. A list of safety-related plant systems and structures was initially identified from the existing components list in the CRL which is part of the plant information management system. The CRL safety classification field was studied to ensure that any plant system and structure with a safety-related component had been considered for inclusion within the scope of license renewal. Additionally the CRL safety classification and associated plant system drawings provided starting points for identifying specific components required to meet the 10 CFR 54.4(a)(1) criterion. During the audit, the applicant described the process for evaluating components classified as safety-related that performed no safety-related intended functions. The applicant stated that the safety classification of several components was reevaluated to reconcile differences between scoping determinations and facility database or CLB information. Identified safety-related components that performed no intended functions and the rationales for their exclusion from scope of license renewal were explicitly described in PP-02. Examples included the containment leak rate testing system, drywell cooling system, and service air system.

The staff reviewed the safety classification criteria to verify consistency between the CLB definition and the Rule definition and reviewed the applicant's evaluation of the differences between the Rule definition and the site-specific definition of "safety-related" to ensure that all potential 10 CFR 54.4(a)(1) SSCs had been adequately addressed. The applicant documented its evaluation in PP-02, stating that the site-specific definition of "safety-related" was nearly identical to the Rule definition with the following three exceptions.

(1)   The CLB defines a safety-related SSC as designed to remain functional for all design basis conditions whereas the Rule defines it as designed to remain functional for all DBEs.

(2)   The CLB definition requires that the reactor be shut down and maintained in a safe (hot) shutdown condition whereas the Rule definition requires that the reactor be maintained in a safe shutdown condition.

(3)   The CLB definition refers to potential 10 CFR Part 100 off-site exposure limits whereas the Rule definition refers also to comparable guidelines in 10 CFR 50.34(a)(1) and 10 CFR 50.67(b)(2).

As to the first exception the staff questioned how non-accident DBEs, particularly those that may not be described in the UFSAR, had been considered during scoping. The applicant responded by identifying applicable DBEs, including external hazards like fire, earthquakes, flooding, wind and missiles, and high-energy line breaks. The additional DBEs were evaluated in PP-13, prepared by the applicant as a primary source for identifying structures and systems within the scope of license renewal. The staff reviewed PP-13, discussed it with the applicant, and finds it a concise and detailed evaluation of these events, including appropriate references to CLB documentation supporting the evaluation, and of systems and structures relied upon to remain functional during and following DBEs. The staff concludes that the applicant has considered a scope of DBEs consistent with SRP-LR guidance.

As to the second exception the applicant verified that all SSCs required to shut down the reactor and maintain it in a cold shutdown condition were considered safety-related at the facility and included within the scope of license renewal in accordance with 10 CFR 54.4(a)(1).

As to the third exception the applicant verified that the comparable guidelines of the cited regulations did not affect the scoping evaluation because the applicant had not revised the current accident source term used in the design basis radiological analysis (10 CFR 50.67(b)(1)) and because 10 CFR 50.34(a)(1)(ii) dose limits pertain only to applicants that applied for construction permits on or after January 10, 1997, which is not the case for OCGS. In addition, the applicant stated that 10 CFR 50.34(a)(1)(i) refers to 10 CFR Part 100 only, as does the CLB.

The staff reviewed the applicant's evaluation and discussed it with the applicant's license renewal team. The staff concludes that the differences between the applicant's "safety-related" definition and the Rule definition had been adequately evaluated by the applicant and had not caused any additional components to be considered safety-related beyond those identified in the CLB.

2.1.4.1.3 Conclusion

Based on this sample review, discussions with the applicant, and review of the applicant's scoping process, the staff concludes that the applicant's methodology for identifying systems and structures meets 10 CFR 54.4(a)(1) scoping criteria and is, therefore, acceptable.

2.1.4.2 Application of the Scoping Criteria in 10 CFR 54.4(a)(2)

2.1.4.2.1 Summary of Technical Information in the Application

In LRA Section 2.1.5.2, "Nonsafety-related affecting safety-related - 10 CFR 54.4(a)(2)," the applicant described the scoping methodology for 10 CFR 54.4(a)(2) nonsafety-related criteria. The applicant evaluated SSCs under 10 CFR 54.4(a)(2) with four categories. The following is a summary description of the four categories:

(1) Nonsafety-related SSCs required for functions that support safety-related system intended functions. The nonsafety-related SSCs credited in the CLB that support safety-related system intended functions were included within the scope of license renewal under 10 CFR 54.4(a)(2) and the scoping evaluation for each system was documented. When a system was included within the scope of license renewal pursuant to 10 CFR 54.4(a)(1), the scoping evaluation included the identification of any additional systems required to support the safety-related system intended function(s).

(2) Nonsafety-related systems connected to and providing structural support for safety-related SSCs. Nonsafety-related systems connected to safety-related systems were entirely within the scope of license renewal under 10 CFR 54.4(a)(2) up to and including the first seismic anchor past the safety-related and nonsafety-related interface, up to a flexible hose or joint not capable of load transfer, or up to the end of the piping run. An anchor or three mutually perpendicular restraints as described in the CLB were considered equivalent to a seismic anchor. Grouted walls or slab penetrations or such anchored components as pumps, heat exchangers, or turbines were also considered equivalent to seismic anchors. Underground piping was also considered equivalent.

(3) Nonsafety-related systems with a potential for spatial interaction with safety-related SSCs. Nonsafety-related systems not directly connected to safety-related piping or components or connected downstream from the first seismic or equivalent anchors were within the scope of license renewal pursuant to 10 CFR 54.4(a)(2) if their failure could adversely impact the performance of safety-related SSC intended functions. Failures considered included nonsafety-related piping failures on adjacent SSCs (e.g., pipe whip, jet impingement, spray, flooding, etc.) and loss of nonsafety-related piping supports causing piping to fall on safety-related SSCs (seismic II/I). To determine which nonsafety-related SSCs were within the scope of license renewal, the applicant evaluated two options, mitigative or preventive.

The mitigative option considered the failure of nonsafety-related systems on safety-related SSCs with the effects controlled by some feature(e.g. whip restraints, spray shields, supports, barriers, etc) installed on the safety-related SSCs. With this mitigation the failure of the nonsafety-related system will not prevent the performance of a 10 CFR 54.4(a)(1) safety-related system intended function. With the mitigative option the mitigative feature (whip restraints, spray shields, supports, barriers, etc.) is included within the scope of license renewal pursuant to 10 CFR 54.4(a)(2). The nonsafety-related systems can be excluded from the scope of license renewal provided the mitigative features are adequate to address all potential failure locations that could result from aging.

For the preventive option, vulnerable safety-related systems in proximity to the nonsafety-related systems are identified by plant walkdowns to identify nonsafety-related systems or portions with the potential for spatial interaction (pipe whip, spray, flooding, etc.) with safety-related equipment, assuming a failure anywhere along the length of the safety-related system. Nonsafety-related SSCs also include heavy load-lifting equipment that could drop on and damage safety-related equipment.

The applicant applied the preventive option for 10 CFR 50.54(a)(2) scoping without consideration of mitigative features. However, certain mitigative features of the CLB were also included within the scope of license renewal. Nonsafety-related systems that contain water, oil, or steam located inside structures with safety-related systems were included within the scope of license renewal for potential spatial interaction under 10 CFR 54.4(a)(2). All supports for nonsafety-related systems with a potential for spatial interaction with safety-related SSCs were included within the scope of license renewal as commodities.

(4) Certain nonsafety-related mitigative plant design features that were part of the CLB. Nonsafety-related SSCs identified as mitigative plant design features in the CLB included turbine building walls (missile protection), walls, dikes, curbs, seals (flood protection), and spray shields.

Air and gas systems were not included within the scope of license renewal under 10 CFR 54.4(a)(2) scoping criteria because they are not hazards to other plant equipment. Plant-specific operating experience verified that they have not adversely affected other plant equipment. Industry operating experience also reveals no events of this nature. Therefore, the applicant concluded that the air/gas systems are not within the scope of license renewal under 10 CFR 54.4(a)(2) scoping criteria. However, supports for air/gas systems with a potential to fall on safety-related systems were included within the scope of license renewal as commodities.

2.1.4.2.2 Staff Evaluation

Pursuant to 10 CFR 54.4(a)(2), the applicant must consider all nonsafety-related SSCs the failure of which could prevent satisfactory performance of safety-related SSCs relied upon to remain functional during and following a DBE to ensure (1) the integrity of the reactor coolant pressure boundary, (2) the ability to shut down the reactor and maintain it in a safe shutdown condition, or (3) the capability to prevent or mitigate the consequences of accidents that could cause potential offsite exposures comparable to those of 10 CFR 50.34(a)(1), 10 CFR 50.67(b)(2), or 10 CFR 100.11.

By letters dated December 3, 2001, and March 15, 2002, the NRC issued a staff position to the NEI with expectations for identifying 10 CFR 54.4(a)(2) SSCs. The December 3rd letter provides specific examples of operating experience with pipe failure events (summarized in NRC Information Notice (IN) 2001-09, "Main Feedwater System Degradation in Safety Related ASME Code Class 2 Piping Inside the Containment of a Pressurized Water Reactor") and the approaches the NRC considers acceptable to determine which piping systems should be included within the scope of license renewal for 10 CFR 54.4(a)(2). The March 15th letter further described the staff's expectations for the evaluation of non-piping SSCs to determine which additional nonsafety-related SSCs are within the scope of license renewal. The position states that applicants should not consider hypothetical failures but rather should base their evaluation on the plant's CLB, engineering judgement and analyses, and relevant operating experience. The letter further describes operating experience as all documented plant-specific and industry experience that can be used to determine the plausibility of a failure. Documentation would include NRC generic communications and event reports, plant-specific condition reports, such industry reports as safety operational event reports, and engineering evaluations.

The staff reviewed LRA Section 2.1.5.2, PLI-02, PP-01, PP-03, and PP-13 Table 2, "Systems and Structures Credited with Operating (Not for Performance of Section 54.4(a)(1) Function) During and Following Non DBA DBEs." PP-01 identifies systems and structures subject to 10 CFR 54.4. PP-01 Attachment 1 lists the 109 systems and 40 structures requiring review for license renewal. PLI-02 describes the process for reviewing these 109 systems and 40 structures and the requirements for entering the results of the review into the license renewal database. The applicable PP, system/structure functions, intended functions, determination of scope of license renewal, supporting systems, and 10 CFR 54.4(a) evaluations were addressed in PLI-02.

The applicant evaluated 10 CFR 54.4(a)(2) SSCs with the four categories taken from the NRC guidance to the industry on identification and treatment of such SSCs:

(1) Nonsafety-related SSCs required for functions that support safety-related SSCs. PLI-2 Sections 6.7 and 6.11, and PP-13 Table 2 implement this process. PLI-2 Section 6.7 provides guidance for identifying support systems. Support systems that support a safety-related system in performing intended functions had to be identified. PLI-2 Section 6.11 required inclusion in the license renewal database scoping input form of the functional support by nonsafety-related SSCs enabling safety-related systems to perform intended functions. PP-13 Table 2 lists nonsafety-related systems not credited with 10 CFR 54.4(a)(1) functions but credited with operating during and following an event. This list was used to determine nonsafety-related systems that support safety-related systems in performing intended functions.

The staff finds that the applicant has implemented an acceptable method for scoping of nonsafety-related systems that perform functions that support safety-related intended functions.

For the remaining three categories, PP-03 provides the criteria for identifying SSCs within the scope of license renewal under 10 CFR 50.54.4(a)(2). PP-03 Section 4.3 states that a spaces approach was used to identify such nonsafety-related SSCs. Initially, structures that house safety-related systems were identified. Structure safety classifications, safety-related system design drawings, and the locations of safety-related components identified in the CRL were used to identify structures that house safety-related components. Seven structures (primary containment, reactor building, emergency diesel generator building, exhaust tunnel, heating boiler house, office building, and turbine building) were identified as containing safety-related systems with components that could fail under wet conditions. These structures, structural components, and component supports were identified as within the scope of license renewal.

Although there are safety-related and nonsafety-related equipment in the miscellaneous yard structure and intake structure, the nonsafety-related equipment in these structures was not included within the scope of license renewal under 10 CFR 50.54(a)(2) because these structures are open to the environment and designed for wet conditions. Nonsafety-related systems in the miscellaneous yard structure are underground with no potential for spacial interaction between safety-related and nonsafety-related systems. The intake structure is classified as safety-related and included within the scope of license renewal pursuant to 10 CFR 50.54(a)(1). Therefore, all intake structural components and component supports were included within the scope of license renewal pursuant to 10 CFR 50.54(a)(1).

All nonsafety-related systems in the primary containment, reactor building, emergency diesel generator building, exhaust tunnel, heating boiler house, office building, and turbine building were evaluated:

(2) Nonsafety-related systems connected to and providing structural support for safety-related SSCs. PLI-02 Section 6.11 and PLI-03 Section 4.5 implement this process. Section 6.11 requires that the establishment of license renewal boundaries between nonsafety-related systems connected to safety-related systems be documented in the license renewal database scoping form. PP-03 Section 4.5 states that the entire nonsafety-related system is within the scope of license renewal under 10 CFR 54.4(a)(2) up to and including the first seismic anchor past the safety-related and nonsafety-related interface, up to a flexible hose or joint not capable of load transfer, or up to the end of the piping run. An anchor or three mutually perpendicular restraints as described in the CLB were considered equivalent to a seismic anchor. Large components like pumps or heat exchangers, piping anchored to walls or slabs, and piping routed underground were also considered equivalent to a seismic anchor. Large components, walls, or slabs were included within the scope of license renewal when credited as seismic anchors.

NEI 95-10 states that an equivalent seismic anchor is typically defined as at least two rigid supports in each of the three orthogonal directions. However, the CLB (Specification 1302-12-294, "Technical Specifications for Oyster Creek Pipe Stress Analysis," Revision 2) states that at least one rigid support in each of the three orthogonal directions is equivalent to a seismic anchor. The staff considered the CLB definition for equivalent seismic anchor in Specification 1302-12-294 appropriate. The staff's review of the LRA, implementing procedures, grouted penetrations, and underground piping identified areas in which additional information was necessary to complete the review of the applicant's scoping and screening results. The applicant responded to the staff's RAIs as discussed below.

PP-03 Section 4.5.1 provides instructions for establishing system boundaries for nonsafety-related piping systems connected directly to safety-related piping systems. One of the acceptable methods in PP-03 for establishing license renewal piping system boundaries is to extend the piping system boundary up to a wall or slab past the safety-related and nonsafety-related interface and credit the grouted wall or slab piping penetration as equivalent to a seismic anchor. The applicant stated that 13 grouted wall or slab piping penetrations were credited as equivalent anchors. Two of the 13 grouted wall or slab piping penetrations were included in stress calculation C-1302-251-5320-004, Revision 4, which demonstrated that these two grouted wall or piping penetrations were equivalent to seismic anchors. No technical analysis demonstrated that the remaining 11 grouted wall or slab piping penetrations were equivalent to seismic anchors.

In RAI 2.1.5.2-1 dated November 9, 2005, the staff requested that the applicant provide technical basis demonstrating that the 11 grouted wall or slab piping penetrations are equivalent to seismic anchors.

In its response dated December 9, 2005, the applicant stated that 7 of the 11 grouted penetrations credited as equivalent to seismic anchors for license renewal had been addressed in the CLB piping analysis. The applicant provided an acceptable technical justification for crediting the remaining 4 grouted piping penetrations as equivalent to seismic anchors in its response. The staff reviewed the applicant's response and concludes that the applicant has adequately described its process for establishing the use of grouted wall penetrations as equivalent to seismic anchors. The staff's concern described in RAI 2.1.5.2-1 is resolved.

LRA Section 2.1.5.2 describes the applicant's screening and scoping methodology for nonsafety-related systems connected to safety-related systems. This section of the LRA states that piping that exits a structure and is routed underground is credited as equivalent to a seismic anchor. This same methodology is described in PP-03 Section 4.5.1.3. During the audit, the applicant clarified that, although described in the LRA and PP-03, this methodology was not used.

In RAI 2.1.5.2-2 dated November 9, 2005, the staff requested that the applicant verify that underground piping was not credited as equivalent to a seismic anchor.

In its response dated December 9, 2005, the applicant stated that underground piping was not credited as an equivalent anchor for license renewal. The staff reviewed the applicant's response and concludes that it has adequately described the process for establishing equivalence to seismic anchors. The staff's concern described in RAI 2.1.5.2-2 is resolved.

(3) Nonsafety-related SSCs not directly connected to safety-related SSCs. PLI-02 Section 6.11 and PP-03 Section 4.6 implement this process. PLI-02 Section 6.11 requires documentation in the license renewal database scoping form of evaluations of any potential adverse interactions between nonsafety-related and safety-related SSCs not physically connected. PLI-03 Section 4.6 states that, although non-liquid systems are not within the scope of license renewal, supports for non-liquid systems in areas of potential seismic interaction with safety-related systems are included. All high-energy lines that contain water, oil, or steam were within the scope of license renewal. All moderate- and low-energy lines that contain water, oil, or steam during plant operation were included within the scope of license renewal. Supports for seismic Class II piping, cranes, monorails, and hoists were also included within the scope of license renewal.

(4) Certain nonsafety-related mitigative plant design features in the CLB. PP-03 Section 4.4 stated that nonsafety-related missile barriers (walls), flood barriers (walls, slabs, curbs, drains, and seals), and spray shields addressed in the CLB are within the scope of license renewal under 10 CFR 54.4(a)(2). Structures with mitigative plant design features were listed in PP-01.

2.1.4.2.3 Conclusion

On the basis of its review and the RAI responses, the staff concludes that the applicant's methodology for identifying systems and structures meets 10 CFR 54.4(a)(2) scoping criteria and is, therefore, acceptable. This determination is based on a review of sample systems, discussions with the applicant, and review of the applicant's scoping process.

2.1.4.3 Application of the Scoping Criteria in 10 CFR 54.4(a)(3)

2.1.4.3.1 Summary of Technical Information in the Application

In LRA Section 2.1.3.4, "Systems and Structures Credited for Regulated Events," the applicant described the scoping methodology for SSCs relied upon in safety analyses or plant evaluation reports performing intended functions. SSCs for fire protection, environmental qualification (EQ), anticipated transient without scram (ATWS), and station blackout (SBO) were included within the scope of license renewal in accordance with 10 CFR 54.4(a)(3). The methodology used to determine the scope of SSCs required by 10 CFR 54.4(a)(3) is described in LRA Section 2.1.3.4. The applicant utilized PPs, PLIs, and the CRL for input to the scoping process.

Fire Protection. In LRA Sections 2.1.3.4, 2.1.4.7, 2.1.6.3, and 2.3.3.15, the applicant described the scoping of SSCs required to demonstrate compliance with 10 CFR 50.48 fire protection requirements. The applicant's technical PP and CLB references for fire protection include (1) PP-07, (2) the Fire Hazard Analysis Report (FHAR), (3) the Fire Safe Shutdown (FSSD) Analysis, (4) UFSAR Section 9.5.1, and (5) the CRL fire protection data field. Using these information sources, the applicant identified components required to support fire protection safe shutdown functions and added them to the license renewal database. SSCs relied upon in safety analyses or plant evaluations to perform functions for compliance with NRC fire protection regulations were included within the scope of license renewal.

Environmental Qualification. In LRA Section 2.1.3.4, the applicant described the scoping of SSCs required to demonstrate compliance with 10 CFR 50.49 EQ requirements. PP-06 summarizes the results of the study of EQ program documents. The applicant selected electrical equipment required for EQ from the EQ Master List. PP-06 lists systems that include EQ components from the EQ Master List of the CRL.

Pressurized Thermal Shock. These requirements are not applicable to OCGS, a boiling water reactor (BWR).

Anticipated Transient Without Scram. In LRA Section 2.1.3.4, the applicant described the scoping of SSCs required to demonstrate compliance with 10 CFR 50.62 ATWS requirements. PP-05 summarizes the CLB as to ATWS and lists systems required by 10 CFR 50.62 to reduce the risk of an ATWS event and structures physically supporting and protecting the credited ATWS systems.

Station Blackout. In LRA Section 2.1.3.4, The applicant described the scoping criteria and in PP-04, the applicant summarizes the CLB as to SBO and lists systems and structures credited with mitigating SBO events.

In accordance with ISG-02, the applicant identified SSCs required to recover from the SBO event and included within the scope of the license renewal. For OCGS, this portion of the plant electrical system connects safety-related buses to onsite emergency power and offsite power to recover from SBO events. Disconnection switches on the supply side of switch yard circuit breakers connecting the 34.5 kV OCGS substation to the plant and continuing through the startup transformers to the switchgear breakers of the plant 4160 alternating current (AC) breakers were included within the scope of license renewal.

2.1.4.3.2 Staff Evaluation

Pursuant to 10 CFR 54.4(a)(3), the applicant must consider all SSCs relied on in safety analyses or plant evaluations to perform functions for compliance with NRC regulations for fire protection, EQ, ATWS, and SBO.

SRP-LR Section 2.1.3.1.3, "Regulated Events," states that all SSCs relied upon in the plant's CLB (as defined in 10 CFR 54.3), plant-specific operating experience, industry operating experience (as appropriate), and safety analyses or plant evaluations to perform functions for compliance with NRC regulations under 10 CFR 54.4(a)(3) must be included within the scope of license renewal. However, hypothetical failures that could result from system interdependencies not part of the CLB and not been previously experienced need not be included.

The staff reviewed the applicant's approach to identifying SSCs relied upon to perform functions related to the four regulated events applicable to BWRs as described in 10 CFR 54.4(a)(3). As part of this review, the staff discussed the methodology with the applicant's license renewal team, reviewed the supporting documentation, and evaluated a sample of the SSCs identified as within the scope of license renewal under 10 CFR 54.4(a)(3).

Fire Protection. For the fire protection regulated event, the staff reviewed the LRA sections noted and PP-07. Components that satisfy fire protection safe shutdown requirements were listed in the FHAR, the FSSD, the CRL fire protection data field, and Appendix R Safe Shutdown Path drawings. The applicant's fire protection confirmation process downloaded CRL fire protection data fields into a database and compared them to FSSD components. This process identified no additional fire protection components.

In addition, LRA Section 2.1.6.3 states in part that equipment stored on site for installation in response to a DBE is considered within the scope of license renewal. The stored equipment credited for 10 CFR Part 50, Appendix R, repairs includes cables and connectors, hoses, tubing, fittings, screws, butts, washers, exhaust fans, and flexible duct. These components were within the scope of license renewal. Tools and supplies used to place stored equipment in service were not within the scope of license renewal. The staff finds the LRA identification of stored equipment within the scope of license renewal acceptable.

In PP-07, Table 1, "Systems Credited for FSSD with Associated FSSD Functions," the applicant listed all FSSD components. In PP-07, Table 2, "Systems Credited for Fire Detection and Suppression," the applicant listed from UFSAR Section 9.5.1 systems required for fire detection and suppression. PP-07, Table 3, "Additional Systems Credited in Commitments Made in Response to Appendix A to Branch Technical Position (BTP) APSCB 9.5-1," the applicant identified additional commitments for systems and components that remove smoke and water and prevent water damage after a fire. The applicant consolidated the three PP-07 tables in Table 4, "Consolidated Table of Systems Relied Upon to Demonstrate Compliance with 10 CFR 50.48." In addition, PP-07, Table 5, "Structures Required to Demonstrate Compliance with 10 CFR 50.48," lists structures and structural support components that comply with fire protection requirements. In the LRA, the applicant used the last two tables to consolidate the scoping effort at the structure and system level.

The staff's review of the LRA identified an area in which additional information was necessary to complete the review of the applicant's scoping and screening results. The applicant responded to the staff's RAIs as discussed below.

PP-07 Section 4 states that first-level, primary support systems necessary for equipment credited in the FHAR or safe shutdown analysis to function for compliance with 10 CFR 54.48 are included within the scope of license renewal. PP-07 Table 1 lists the standby gas engine (propane) generator as within the scope of license renewal. However, LRA Section 2.5.1.15 does not list the backup gas (propane) engine generator as within the scope of license renewal. The applicant stated during the audit that LRA Section 2.5.1.15 is correct and that the backup gas (propane) generator was removed from the scope of license renewal because it is not the radio communication system's primary power source.

In RAI 2.5.1.15-1 dated November 9, 2005, the staff requested that the applicant:

(1) Verify that the CLB, plant-specific experience, industry experience (as appropriate), and safety analyses or plant evaluations do not require the backup gas (propane) generator to perform a function for compliance with NRC regulations under 10 CFR 54.4(a)(3).

(2) Verify that second-, third-, or fourth-level support systems were included within the scope of license renewal if the CLB, plant-specific experience, industry experience (as appropriate), and safety analyses or plant evaluations require such support systems to perform functions for compliance with NRC regulations under 10 CFR 54.4(a)(3).

In its responses dated December 9, 2005, and June 7, 2006, the applicant stated that it had determined that the repeater located at the Meteorological Tower (Met Tower) is credited for communication capabilities for some 10 CFR Part 50, Appendix R, scenarios. Therefore, the repeater and associated support equipment, including the backup gas (propane) engine generator located at the Met Tower, are now within the scope of license renewal and subject to an AMR. The applicant also stated that the second-, third-, and fourth-level support systems were included within the scope of license renewal if the CLB, plant-specific experience, industry experience, and safety analyses or plant evaluations require these systems to perform functions for compliance with 10 CFR 54.4(a)(3). The staff reviewed the applicant's response and concludes that it is adequate. The staff's concerns described in RAI 2.5.1.15-1 are resolved.

Based on the review of the LRA, PP-07, and ISGs the staff finds that the fire protection implementing documents for license renewal meet 10 CFR 54.4(a)(3) requirements.

Environmental Qualification. For the EQ regulated event, the staff evaluated LRA Section 2.1.3.4 and PP-06. The UFSAR Section 3.11.1.1.1, "Criteria for Selection of Equipment," identifies the scope of electrical equipment and components that must be environmentally qualified for use in harsh environments. The electrical components in the EQ Master List were entered into the CRL, which CRL includes an EQ data field for identifying EQ components. In PP-06 Table 1, "Systems Subject to 10 CFR 50.49 EQ Requirements," the applicant identified mechanical, electrical, and instrumentation and control (I&C) systems with EQ equipment within the scope of license renewal. PP-06 Table 1 was compared to the EQ Master List to verify that the EQ Master List was consistent with the CRL. In PP-06 Table 2, "Structures Associated with EQ

Environmental Boundaries," the applicant identified structures that provide physical boundaries for postulated harsh environments with EQ electrical equipment included within the scope of license renewal: the containment, reactor building, turbine building, standby gas treatment exhaust tunnel, containment electrical penetrations, and EQ barriers in the 4160V switchgear.

The staff finds that the LRA and PP-06 adequately identified the scope of EQ electrical systems, electrical penetrations, cable routing and terminations, and structures within the scope of license renewal.

Anticipated Transient Without Scram. For the ATWS regulated event, the staff evaluated LRA Section 2.1.3.4 and PP-05. PP-05, Attachment 1, identifies systems within the scope of license renewal. PP-05, Attachment 2, identifies the primary containment, reactor building, turbine building, and the component supports commodity group as within the scope of license renewal. The staff finds that the LRA and PP-05 adequately identify ATWS SSCs within the scope of license renewal.

Station Blackout. For the SBO regulated event, the staff evaluated LRA Sections 2.1.3.4 and 2.1.4 and several mechanical, structural, and electrical systems in LRA Sections 2.3, 2.4, and 2.5. The staff compared the LRA information to that of PP–04, Table I, "Systems and Structures Credited to Cope with an SBO Event," Table II, "Systems Credited for Safe Shutdown During a Station Blackout," Table III, "Systems Required to Recover from a Station Blackout Event," and Table IV, "Structures Required For Station Blackout Event," where the applicant identified the SBO electrical and mechanical systems and components and support structures that house SBO equipment within the scope of license renewal needed under 10 CFR 54.4(a)(3) to meet the SBO regulated event.

In PP-04, the applicant stated that it had added the alternate AC (AAC) power supply system to the existing plant configuration to comply with the SBO rule. The AAC source is provided by one of two non-Class IE combustion turbines located at the Forked River site adjacent to OCGS. The AAC source supplies power to OCGS via a connection to the non-1E 4160V "1B" switchgear. In PP-04, Table II, the AAC combustion turbines and their sub-systems, the turbine lube oil system, the fuel system, the direct current (DC) power system, and the SBO transformer are parts of the AAC Power Supply System within the scope of license renewal for the SBO regulated event under 10 CFR 54.4(a)(3). In PP-04, Table IV, the applicant identified the Forked River Combustion Turbine (FRCT) buildings as support structures protecting relay cables, I&C cables, combustion turbines, and other equipment.

In LRA Table 2.5.1.19, the ACC combustion turbines are identified as one combustion turbine power plant unit within the scope of license renewal and subject to an AMR. As described in SER Section 2.5.5.2, in its response to RAI 2.5.1.19-1, the applicant stated that it had revised the combustion turbine power plant unit scoping and screening methodology. Mechanical, electrical, and structural component types were itemized in detail consistent with scoping and screening methodology for other license renewal systems and structures.

The staff finds that the LRA, as revised in the response to RAI 2.5.1.19-1, and the methodology as described in PP-04 has adequately identified SSCs within the scope of license renewal for the SBO regulated event.

2.1.4.3.3 Conclusion

Based on the sample review, RAI responses, discussions with the applicant, and review of the applicant's scoping process, the staff concludes that the applicant's methodology for identifying systems and structures meets 10 CFR 54.4(a)(3) scoping criteria and is, therefore, acceptable.

2.1.4.4 Plant-Level Scoping of Systems and Structures

2.1.4.4.1 Summary of Technical Information in the Application

System and Structure Level Scoping. In LRA Section 2.1, the applicant described the scoping methodology for safety-related and nonsafety-related systems and structures and equipment relied upon for functions for 10 CFR 54.4(a)(3) regulated events. The scoping methodology is consistent with guidance by the NRC in the SRP-LR and by the industry in NEI 95-10. In LRA Section 2.2, using the methodology described in LRA Section 2.1, the applicant evaluated systems and structures to determine whether they were within the scope of license renewal. The results of plant scoping are provided in LRA Table 2.2-1.

Component Level Scoping. The applicant identified the systems and structures within the scope of license renewal and determined the components within each mechanical system and structure. The structural and mechanical components supporting intended functions were considered within the scope of license renewal and screened to determine whether AMRs were required. All electrical components of in-scope mechanical and electrical systems were included as commodity groups. The applicant considered three component classifications during this stage of the scoping methodology: mechanical, structural, and electrical. The CRL lists plant mechanical components comprehensively. The database identifies components by type and unique number. In the scoping and screening results section of the LRA (Sections 2.3, 2.4 and 2.5), components are identified by component type only.

Commodity Groups Scoping. All electrical components of in-scope of mechanical and electrical systems were included as commodity groups. Many active electrical commodity groups were screened out and not subject to an AMR. In LRA Section 2.5.2, the applicant described the commodity groups used to evaluate all in-scope electrical components subject to an AMR.

Structural components were grouped as component types based on design function, materials of construction, and environments. LRA Section 2.4 states that such component types as component supports and piping and component insulation were placed in commodity groups.

Insulation. LRA Section 2.4.19 states that insulation installed on hot piping or components of structures within the scope of license renewal (with the exception of miscellaneous yard structures) were included within the scope of license renewal as a commodity group. All insulation was considered nonsafety-related. Therefore, the piping and component insulation commodity group is within the scope of license renewal under 10 CFR 54.4(a)(2) because insulation performs a function that supports a 10 CFR 54.4(a)(1) system. Piping and component insulation in the miscellaneous yard structure is not within the scope of license renewal because its failure does not impact any safety-related intended function.

Consumables. LRA Section 2.1.6.4, the applicant discussed consumables, using the guidance in SRP-LR Table 2.1-3 to categorize and evaluate consumables. Consumables were divided into the following four categories for the purpose of license renewal: (a) packing, gaskets, component seals, and o-rings, (b) structural sealants,(c) oil, grease, and component filters, and (d) system filters, fire extinguishers, fire hoses, and air packs.

Group (a) subcomponents are not relied on to form a pressure-retaining function and, therefore, are not subject to an AMR. Group (b) structural sealants for structures within the scope of license renewal require an AMR. Group(c) subcomponents are periodically replaced in accordance with plant procedures and therefore are not subject to an AMR. Group (d) consumables are subject to replacement based on National Fire Protection Association standards in accordance with plant procedures and, therefore, are not subject to an AMR.

2.1.4.4.2 Staff Evaluation

The staff reviewed the applicant's methodology for scoping plant systems and components for consistency with 10 CFR 54.4(a). The methodology used to determine the systems and components within the scope of license renewal is documented in PP-01, PP-02, PP-04, PP-05, PP-06, PP-07, PP-08, PP-13, and PLI-02, and plant level scoping results are identified in LRA Table 2.2-1. The scoping process defined the entire plant in terms of systems and structures. Specifically, PP-01 identifies systems and structures subject to 10 CFR 54.4 review. PLI-02 describes the process for entering process results into the license renewal database. PP-02 and PP-13 were used to determine whether the system or structure was safety-related. PP-03 was used to determine whether failure of a nonsafety-related system or structure could prevent a safety-related system or structure from performing an intended function. PP-04 (SBO), PP-05 (ATWS), PP-06 (EQ), and PP-07 (fire protection) were used to determine whether the system or structure is relied upon for compliance with NRC regulation of such events. PP-01, PP-03, and PP-08 describe the commodity groups. The process was completed for all systems and structures to ensure that the entire plant was addressed. The applicant's personnel initially evaluated systems and structures identified in the CLB.

The staff noted that a system or structure was presumed to be within the scope of license renewal if it performed one or more safety-related functions or met other scoping criteria pursuant to the Rule as determined by CLB review. Mechanical and structural component types that supported intended functions were considered within the scope of license renewal. All component types in electrical systems within the scope of license renewal were considered within the scope of license renewal and placed in commodity groups. The electrical commodity groups were further screened to determine whether they required AMRs. The staff finds no discrepancies with the methodology used by the applicant.

The staff reviewed the methodology used by the applicant to generate commodity groups. Three separate commodity groups are identified in PP-01 (electrical, component supports, and piping and component insulation). The staff reviewed the commodity group level functions evaluated by the applicant in accordance with 10 CFR 54.4(a). This process determined whether the commodity group had been considered within the scope of license renewal. The staff finds the methodology acceptable.

The staff reviewed the results of the scoping process documented in accordance with PLI-02. This documentation describes the system or structure and its 10 CFR 54.4(a) scoping criteria. The staff also reviewed a sample of the applicant's scoping documentation and concludes that it contains an appropriate level of detail to document the scoping process.

The applicant examined the CLB and determined that insulation installed on hot piping or components in structures within the scope of license renewal (with the exception of miscellaneous yard structures) was included within the scope of license renewal as a commodity subject to an AMR. The staff concludes that the applicant's methods and conclusions as to insulation were acceptable.

The staff reviewed the scoping and screening of consumables and finds that the applicant had followed the process described in the SRP-LR.

2.1.4.4.3 Conclusion

Based on review of the LRA, CRL, scoping and screening implementation procedures, and a sampling of system scoping results during the audit, the staff concludes that the applicant's scoping methodology for plant SSCs, commodity groups, insulation, and consumables is acceptable. In particular, the staff finds that the applicant's methodology reasonably identifies systems, structures, component types, and commodity groups within the scope of license renewal and their intended functions.

2.1.4.5 Mechanical Component Scoping

2.1.4.5.1 Summary of Technical Information in the Application

In LRA Sections 2.1.5.5 and 2.3.1, the applicant discussed the scoping methodology for mechanical systems and components. For mechanical systems, mechanical components supporting system intended functions are included within the scope of license renewal. Mechanical system diagrams are marked to create LRBDs showing in-scope components that support safety-related functions or regulated events highlighted in green; nonsafety-related components connected to safety-related components and providing structural support at the connections or components the failure of which could prevent satisfactory accomplishment of a safety-related function due to spatial interaction with safety-related SSCs are highlighted in red. A computer sort from the CRL was compared against the LRBDs to confirm the scope of components in the system. For additional information, the applicant performed plant walkdowns when required.

2.1.4.5.2 Staff Evaluation

The staff evaluated LRA Sections 2.1.5.5 and 2.3.1 and the guidance in PLI-02 and PLI-04 to complete the review of the mechanical scoping process. PLI-04 utilizes information in PP-01 through PP-07 to complete the mechanical scoping process.

PLI-2 provides instructions for filling out system data fields in the license renewal database. The license renewal database was used to develop license renewal system and structure scoping forms for subsequent review, approval, and document retention. The CLB documents were utilized when determining whether a system or component was within the scope of 10 CFR 54.4(a). The CLB includes the UFSAR, the facility description safety analysis report, separate ATWS, EQ, fire protection, and SBO documents, technical specifications, SERs, the Integrated Plant Safety Assessment Report, and NRC orders. Other documents included the CRL, flow diagrams, licensed operator training plans, and the Maintenance Rule database. In the event of differences between CLB documents and other documents, the CLB documents took precedence.

The license renewal database scoping input forms included the following information: license renewal system name, system grouping, DBD if applicable, UFSAR sections, drawings, other reference documents, and system intended functions. The applicant then evaluated the 10 CFR 54.4(a) scoping criteria against the identified system intended functions to determine which criteria applied. The applicant also identified support system intended functions which provide the functional and physical support required to accomplish safety-related intended functions. Using PLI-04, the applicant then created LRBDs for mechanical systems.

The staff finds the PLIs and PPs acceptable in identifying mechanical components and support structures in mechanical systems within the scope of license renewal.

Scoping Methodology for the Isolation Condenser System. In LRA Section 2.3.1.3, the applicant provided the scoping and screening methodology results for SSCs within the ICS. The ICS is a safety-related system credited with mitigating the effects of feedwater loss and specific high-energy line breaks. The ICS license renewal scoping boundary includes those portions of nonsafety-related piping and equipment extending beyond the safety-related and nonsafety-related interface. The scoping results indicated that the ICS contains seven system functions within and two system functions not within the scope of license renewal. The staff identified no issues with the ICS scoping results. The staff reviewed the applicant's methodology for identifying ICS mechanical and electrical component types with scoping criteria as defined in the Rule. The staff also reviewed a sample of the scoping methodology implementation procedures and discussed the methodology and results with the applicant. The staff verified that the applicant had used pertinent engineering and licensing information to identify the ICS mechanical, structural, and electrical component types within the scope of license renewal.

2.1.4.5.3 Conclusion

Based on the staff's review of the information in the LRA, PLIs, PPs, and the system sample and discussions with the applicant, the staff concludes that the applicant's methodology for identifying mechanical systems for 10 CFR 54.4(a) scoping criteria is acceptable.

2.1.4.6 Structural Component Scoping

2.1.4.6.1 Summary of Technical Information in the Application

In LRA Section 2.1 the applicant described the methodology used for structural scoping. Additional details of the scoping methodology for structures is provided in PP-01, PP-02, PP-03, PP-013, and PLI-02. Following the initial identification of all structures, the applicant identified intended functions as the bases for including specific structures within the scope of license renewal. The structure intended functions are based on applicable CLB reference documents. The applicant then identified all structural components that support the intended functions and included them within the scope of license renewal as component types. The structural components were identified from a review of applicable plant design drawings of the structure and supplemental plant walkdowns when required for additional confirmation. A single site plan layout drawing was marked up to create an LRBD showing in-scope structures.

2.1.4.6.2 Staff Evaluation

Structural scoping ensured that all plant buildings, yard structures and their constituent parts were considered for license renewal. Initially PP-01 was prepared to establish a comprehensive list of license renewal structures and to document the basis for the list. The structures list was then compared to the CRL, including the UFSAR, plant design drawings, the maintenance rule database, and other plant design documents to ensure that it was comprehensive and consistent with the CLB. The resultant list of structures was categorized as "Structures and Component Supports" for further evaluation.

Following identification of all plant structures, the applicant implemented PLI-02 to evaluate them, identify their functions, and determine which are intended functions required for compliance with one or more 10 CFR 54.4(a) criteria. Various other PPs (PP-02 through PP-07) were developed to support the evaluation of each structure in accordance with the scoping criteria. For each structure, the applicant further studied the drawings and plant databases to identify specific structural components and features. The structural component intended functions were identified based on the guidance of Regulatory Guide 1.188, "Standard Format and Content for Applications to Renew Nuclear Power Plant Operating Licenses," NEI 95-10, and the SRP-LR. Procedures also described the source design documentation used for the evaluation of structures including the various technical PPs developed by the applicant to support the LRA. For structures, the evaluation boundaries were determined from a complete description of each structure according to intended functions performed and its components per PLI-04. The license renewal database was used to compile the structural evaluation results. The database contains a list of structures, structural component types, evaluation results for each of the 10 CFR 54.4(a) criteria for each structure, a description of structural intended functions and source reference information for the functions, and a reference to pertinent plant layout drawing(s) for each structure. Plant structures within the scope of license renewal were captured on a plant layout drawing. The boundaries of the structures were identified from the physical representation of the structure on the layout drawing.

The staff conducted detailed discussions with the applicant's license renewal team and reviewed documentation pertinent to the scoping process. The staff assessed whether the scoping methodology and procedures outlined in the LRA had been appropriately implemented and whether the scoping results were consistent with CLB requirements. The staff also reviewed structural scoping evaluation results for the reactor building for proper implementation of the scoping process for structural components and compared a sample of structural components identified in the reactor building structural drawings to the structural list in the license renewal database for consistency. In these audit activities, the staff identified no discrepancies between the methodology documented and the implementation results.

2.1.4.6.3 Conclusion

Based on review of information in the LRA, the applicant's detailed scoping implementation procedures, and a sampling of structural scoping results, the staff concludes that the applicant's methodology for identification of structural component types within the scope of license renewal meets 10 CFR 54.4(a) requirements and is, therefore, acceptable.

2.1.4.7 Electrical Component Scoping

2.1.4.7.1 Summary of Technical Information in the Application

LRA Sections 2.1.1 and 2.1.5.5 describe the scoping process for electrical systems and components. All electrical systems were evaluated in accordance with 10 CFR 54.4(a) scoping criteria. A system was included within the scope of license renewal if it performed one or more intended functions. The entire system was included within the scope of license renewal if any portion of the system met 10 CFR 54.4(a) scoping criteria. A single electrical boundary drawing was prepared to show schematically portions of the plant electrical distribution system included within the scope of license renewal. The CRL was used to identify electrical components. All electrical components of electrical and mechanical systems within the scope of license renewal were included within the scope of license renewal as commodity groups.

2.1.4.7.2 Staff Evaluation

The staff evaluated LRA Sections 2.1.1 and 2.1.5.5 and implementing procedures PP-01, PP-04, PP-05, PP-06, PP-07, PP-08, and PLI-02. The staff also evaluated the single electrical boundary drawing specifically developed for license renewal showing portions of the plant electrical distribution system included within the scope of license renewal. The staff reviewed the electrical systems and electrical components in mechanical systems identified in the ICS scoping form. The staff discussed the electrical scoping methodology with the applicant's LRA team.

The CRL and UFSAR were used primarily to identify electrical systems and electrical components in mechanical systems within the scope of license renewal. PP-01 identifies the systems within the scope of review for license renewal. PP-04, PP-05, PP-06, and PP-07 specifically identify the electrical and mechanical systems credited for meeting SBO, ATWS, EQ, and fire protection regulatory requirements. The electrical commodity groups are identified in PP-08. PLI-2 provides instructions for filling out system data fields in the license renewal database.

2.1.4.7.3 Conclusion

Based on review of information in the LRA, the applicant's detailed scoping implementation procedures, and a sampling of electrical scoping results, the staff concludes that the applicant's methodology for identification of electrical components within the scope of license renewal meets 10 CFR 54.4(a) requirements, and is, therefore, acceptable.

2.1.4.8 Conclusion for Scoping Methodology

Based on a review of the LRA and the scoping implementation procedures, the staff concludes that the applicant's scoping methodology is consistent with SRP-LR guidance and identified safety-related SSCs the failure of which could affect safety-related functions and which are necessary for compliance with the NRC's regulations for fire protection, EQ, ATWS, and SBO. Therefore, the staff concludes that the applicant's methodology meets 10 CFR 54.4(a) requirements.

2.1.5 Screening Methodology

2.1.5.1 General Screening Methodology

After identifying systems and structures within the scope of license renewal, the applicant implemented a process for identifying SCs subject to an AMR, in accordance 10 CFR 54.21.

2.1.5.1.1 Summary of Technical Information in the Application

In LRA Section 2.1.6, the applicant discussed the method of identifying components of in-scope systems and structures subject to an AMR. The identification method consisted of the following steps:

(1) Identification of long-lived and passive components for each in-scope mechanical system, structure, and electrical commodity group.

(2) Identification of the license renewal intended function(s) for all mechanical and structural component types and electrical commodity groups.

Active components were screened out and required no AMR. The screening process also identified short-lived components and consumables. Short-lived components are not subject to an AMR. Consumables are a special class that includes packing, gaskets, component seals, o-rings, oil, grease, component filters, system filters, fire extinguishers, fire hoses, and air packs. Structural sealants were the only consumables within the scope of license requiring an AMR.

2.1.5.1.2 Staff Evaluation

Pursuant to 10 CFR 54.21, each LRA must contain an IPA that identifies SCs within the scope of license renewal and subject to an AMR. The IPA must identify components that perform intended functions without moving parts or a change in configuration or properties (passive) as well as components not subject to periodic replacement based on a qualified life or specified time period (long-lived). The IPA includes a description and justification of the methodology used to identify passive and long-lived SCs and a demonstration that the effects of aging on those SCs will be adequately managed so that intended function(s) will be maintained under all design conditions imposed by the CLB for the period of extended operation.

The staff reviewed the methodology used by the applicant to determine whether mechanical and structural component types and electrical commodity groups within the scope of license renewal should be subject to an AMR. The applicant implemented a process for determining which SCs were subject to an AMR in accordance with 10 CFR 54.21(a)(1) requirements. In LRA Section 2.1.6, the applicant discussed screening of component types and commodity groups within the scope of license renewal.

The screening process evaluated these in-scope component types and commodity groups to determine which were long-lived and passive and, therefore, subject to an AMR. The staff reviewed LRA Sections 2.3, 2.4, and 2.5 that provide the results of the process used to identify component types and commodity groups subject to an AMR. The staff also reviewed the screening results reports for the ICS and the reactor building.

The applicant discussed with the staff in detail the processes for each discipline and provided administrative documentation that described the screening methodology. Specific methodology for mechanical, electrical, and structural is discussed below.

2.1.5.1.3 Conclusion

On the basis of review of the LRA, the screening implementation procedures, and a sampling of screening results, the staff concludes that the applicant's screening methodology is consistent with SRP-LR guidance and capable of identifying passive, long-lived components within the scope of license renewal and subject to an AMR. The staff finds that the applicant's process for identifying component types and commodity groups subject to an AMR meets 10 CFR 54.21 requirements and is, therefore, acceptable.

2.1.5.2 *Mechanical Component Screening*

2.1.5.2.1 Technical Information in the Application

In LRA Section 2.1.6.1, the applicant discussed the screening methodology for identifying passive and long-lived mechanical components and their support structures subject to an AMR. The mechanical system screening process began with the results from the scoping process. The applicant studied LRBDs to identify passive and long-lived components, then entered them into the license renewal database. The applicant also examined components in the CRL to confirm that all system components had been considered. Where the LRBDs did not provide sufficient detail, as for large vendor-supplied components (e.g., compressors, emergency diesel generators), the applicant examined associated component drawings or vendor manuals. The applicant also performed plant walkdowns to confirm which components required an AMR. Finally, the applicant benchmarked passive and long-lived components for a system against previous LRAs with similar systems.

2.1.5.2.2 Staff Evaluation

The staff evaluated the mechanical screening methodology in LRA Section 2.1.6.1, PLI-03, and PP-08. Using PLI-03 for mechanical systems, the applicant downloaded a listing of components from the CRL to assist in identifying system passive, long-lived component types.

An important function in the screening form is the "Intended Function" column. The list of potential intended functions is identified in PP-08 and included in the pull-down menu for the intended functions database field. For components like restricting orifices or heat exchangers, the appropriate intended function depends on the specific application within the system or structure. For example, the in-scope heat exchanger has a pressure boundary intended function, but the tubes have a heat transfer function if required to support a system intended function under 10 CFR 54.4(a). All in-scope passive, long-lived mechanical components have at least one intended function.

Based on the mechanical screening methodology in LRA Section 2.1.6.1, PLI-03, and PP-08, the staff finds the mechanical screening process acceptable.

Screening Methodology for the Isolation Condenser System. In LRA Table 2.3.1.3, the applicant identified the following isolation condenser system component types and intended functions subject to an AMR:

- bird screen - filter
- closure bolting - mechanical closure
- gauge snubbers - pressure boundary
- heat exchangers (isolation condensers) - heat transfer and pressure boundary
- piping and fittings - leakage and pressure boundary
- thermowell - pressure boundary
- valve body - leakage and pressure boundary

The staff questioned the applicant to determine whether instrument lines had been included within the scope of license renewal and subject to an AMR. The applicant stated that instrument lines that penetrate the ICS and serve pressure boundary functions were covered under piping and fittings. The ICS and structure screening form lists ICS steam supply instrument lines. The staff also questioned the applicant about expansion joints on the isolation condenser outlet to atmosphere from the isolation condenser heat exchangers. The applicant stated that expansion joints are pipe fittings included within the scope of license renewal and subject to an AMR.

The applicant used PP-08 and PLI-03 to identify the components subject to an AMR.

2.1.5.2.3 Conclusion

Based on a review of the LRA, the screening implementation procedures, and a sample of isolation condenser system screening results, the staff concludes that the applicant's mechanical component screening methodology is consistent with SRP-LR guidance. The staff concludes that the applicant's methodology for identification of mechanical components subject to an AMR meets 10 CFR 54.21(a)(1) requirements.

2.1.5.3 Structural Component Screening

2.1.5.3.1 Summary of Technical Information in the Application

The applicant described the methodology for structural screening in LRA Section 2.1.6.1. Additional details related to the implementation of the screening methodology for structures is provided by PP-08 and PLI-03. The applicant's structure screening process began with the results from the scoping process. For all in-scope structures, the applicant reviewed the completed scoping packages, which included written descriptions of each structure or structure portion as well as the structure drawings to identify the passive, long-lived SCs. The SRP-LR and NEI 95-10 Appendix B were used to identify passive SCs. These were then entered into the license renewal database and the component listings compared against the CRL to confirm that all structural components had been considered. Plant walkdowns were performed when required for confirmation. Finally, the list of identified passive, long-lived SCs was benchmarked against previous LRAs. Components which support or interface with electrical components, for example, cable trays, conduits, instrument racks, panels and enclosures, were assessed as structural components.

2.1.5.3.2 Staff Evaluation

The staff reviewed the applicant's methodology for structural screening described in LRA Section 2.1.6.1 and in implementing guidance in PP-08 and PLI-03. The scoping results show that the applicant screened per the PLI and used the screening data forms within the license renewal database to capture pertinent structure design information, component or commodity types, materials, environments, and aging effects. As to the component type, the staff verified that the applicant had used the lists of passive SCs embodied in the regulatory guidance as a starting point and supplemented that list with additional items unique to the site or for which a direct match to the generic lists did not exist (i.e., material/environment combinations). As one of the general rules for structural screening, the applicant determined that components which support or interface with electrical components, (e.g., cable trays, conduits, instrument racks, panels and enclosures,) were assessed as structural components.

The staff reviewed the methodology used by the applicant to determine whether structures within the scope of license renewal would be subject to further AMR. For structures, the applicant determined the types of structural elements utilized and the various materials and environments to be considered in the AMR. Generally, the boundary for a structure is the entire building including base slabs, foundations, walls, beams, slabs, and steel superstructure. A listing of all the systems and component types in each plant structure was developed identifying the various structural elements, materials, and environments. The applicant created a database to compile the results. The database identified each SC and indicated whether the component type was subject to an AMR. Each component type was identified as a component (e.g., door, gate, anchor support, strut, fastener, or siding) or as a material (e.g., concrete, polymer, or steel). From this identification a screening report for each plant structure was developed. The applicant described and discussed with the staff in detail the screening methodology as well as the screening reports for a selected group of structures.

The staff reviewed the applicant's results from the implementation of this methodology for one of the plant structures (reactor building) identified as within the scope of license renewal. The staff also reviewed the various reactor building structural drawings to verify that the applicant had performed a comprehensive evaluation and had identified the relevant structures and structural elements in the evaluation. The review included in-scope components, the corresponding component-level intended functions, and the resulting list of component types subject to an AMR. The staff also discussed the process and its results with the applicant. The staff identified no discrepancies between the methodology documented and the implementation results.

2.1.5.3.3 Conclusion

Based on review of information in the LRA, the applicant's detailed screening implementation procedures, and a sampling of structural screening results, the staff concludes that the applicant's methodology for identification of structural component types subject to an AMR meets 10 CFR 54.21(a)(1) requirements.

2.1.5.4 Electrical Component Screening

2.1.5.4.1 Summary of Technical Information in the Application

In LRA Sections 2.1.6 and 2.5.2, the applicant discussed the method for identifying electrical components in systems within the scope of license renewal. Initially, electrical component types in the electrical and mechanical systems within the scope of license renewal were identified. Component types from drawings and the CRL were grouped into approximately 52 electrical commodity groups based on guidance in NEI 95-10 Appendix B and NUREG-1801, Revision 1, "Generic Aging Lessons Learned (GALL) Report," dated September 2005. Forty of the commodity groups were classified as active and therefore not subject to an AMR. Two of the remaining twelve commodity groups were not subject to an AMR because they performed no license renewal intended functions. Components in the EQ program replaced prior to expiration of their qualified lives were screened out from requiring an AMR. The remaining eight commodity groups were subject to an AMR. Insulated cables and connections, electrical penetrations, high voltage insulators, transmission conductors and connections, fuse holders, wooden utility poles, cable connections (metallic parts), and uninsulated ground conductors were the commodity groups identified by the applicant in the LRA subject as to an AMR.

In its response to RAI 2.5.1.19-1, the applicant stated that it had revised its approach to aging management for the SBO combustion turbine power plant. Table 2.5.2A of the RAI response identifies nine SBO electrical commodity groups. Cable connections (metallic parts), high voltage insulators, insulated cables and connections, insulated inaccessible medium-voltage cables, phase bus connections, phase bus enclosure assemblies, phase bus insulators, transmission conductor and connections, and uninsulated ground conductors were identified as commodity groups in the RAI response.

2.1.5.4.2 Staff Evaluation

The staff evaluated the applicant's methodology for electrical screening in LRA Sections 2.1.6 and 2.5.2, PP-08, and the response to RAI 2.5.1.19-1. The applicant used the screening process described in PP-08, PLI-02, and the RAI response to identify the electrical commodity groups subject to an AMR. Components types within electrical systems determined to require an AMR were placed in commodity groups. The commodity groups established for passive, long-lived component types were evaluated to determine whether they were subject to replacement based on a qualified life or specified time period (short-lived) or not (long-lived).

The applicant stated in the LRA that most electrical commodity groups were active. Using NEI 95-10 Appendix B as guidance, the applicant screened out active commodity groups as not requiring an AMR pursuant to 10 CFR Part 54.

The staff reviewed the applicant's approach to scoping and screening of electrical fuse holders in accordance with ISG-05, "Identification and Treatment of Electrical Fuse Holders for License Renewal," which states that, consistent with 10 CFR 54.4(a) specified requirements, fuse holders (including fuse clips and fuse blocks) are considered passive electrical components. Fuse holders should be scoped, screened, and included in the AMR in the same manner as terminal blocks and other types of electrical connections treated in the process. ISG-05 also states that fuse holders of an active component assembly (i.e., switchgear, power supplies, power inverters, battery chargers and circuit boards) are not subject to an AMR.

The staff reviewed and discussed the applicant's evaluations of fuse holders. The applicant examined fuse holders not included in the EQ program or inside active equipment and determined that such fuse holders were subject to an AMR.

2.1.5.4.3 Conclusion

The staff reviewed the LRA, procedures, license renewal electrical schematic, a sample of the results of the screening methodology, and the applicant's response to RAI 2.5.1.19-1. The staff concludes that the applicant's methodology is consistent with the description in LRA and with the applicant's implementing procedures. Based on review of information in the LRA, the applicant's screening implementation procedures, and a sampling of electrical screening results, the staff concludes that the applicant's methodology for identification of electrical commodity groups subject to an AMR meets 10 CFR 54.21(a)(1) requirements.

2.1.5.5 Conclusion for Screening Methodology

After review of the LRA and the screening implementation procedures, discussions with the applicant's staff, and a sample review of screening results, the staff concludes that the applicant's screening methodology is consistent with SRP-LR guidance and has identified those passive, long-lived components within the scope of license renewal and subject to an AMR. The staff concludes that the applicant's methodology meets 10 CFR 54.21(a)(1) requirements and is, therefore, acceptable.

2.1.6 Conclusion for Scoping and Screening Methodology

The staff reviewed the information presented in LRA Section 2.1, the supporting information in the scoping and screening implementation procedures and reports, the information presented during the scoping and screening methodology audit, and the applicant's responses to RAIs 2.5.1.19-1, 2.1.5.2-1, and 2.5.1.2-2. The staff concludes that the applicant's methodology for identifying SSCs within the scope of license renewal and SCs requiring an AMR is consistent with 10 CFR 54.4 and 10 CFR 54.21(a)(1) requirements.

2.2 Plant-Level Scoping Results

2.2.1 Introduction

In LRA Section 2.1, the applicant described the methodology for identifying SSCs within the scope of license renewal. In LRA Section 2.2, the applicant used the scoping methodology to identify SSCs within the scope of license renewal. The staff reviewed the plant-level scoping results to determine whether the applicant had properly identified all plant-level systems and structures relied upon to mitigate DBEs, as required by 10 CFR 54.4(a)(1), or the failure of which could prevent satisfactory performance of any of the safety-related functions, as required by 10 CFR 54.4(a)(2), as well as the systems and structures relied on in safety analysis or plant evaluations for functions required by one of the regulations to which 10 CFR 54.4(a)(3) refers.

2.2.2 Summary of Technical Information in the Application

In LRA Table 2.2-1, the applicant provided a list of the plant systems, structures, and commodity groups evaluated to determine whether they are within the scope of license renewal. Based on the DBEs considered in the plant's CLB, other CLB information on nonsafety-related systems and structures and certain regulated events, the applicant identified those plant-level systems and structures within the scope of license renewal, as defined by 10 CFR 54.4.

2.2.3 Staff Evaluation

In LRA Section 2.1, the applicant described its methodology for identifying systems and structures within the scope of license renewal and subject to an AMR. The staff reviewed the scoping and screening methodology and its evaluation is in SER Section 2.1. To verify that the applicant had properly implemented its methodology, the staff focused its review on the implementation results shown in LRA Table 2.2-1 to confirm that there were no omissions of plant-level systems and structures within the scope of license renewal.

The staff determined whether the applicant had properly identified systems and structures within the scope of license renewal in accordance with 10 CFR 54.4. The staff reviewed selected systems and structures that the applicant had not identified as within the scope of license renewal to verify whether they had any intended functions requiring their inclusion within the scope of license renewal. The staff's review of the applicant's implementation was conducted in accordance with the guidance of SRP-LR Section 2.2, "Plant-Level Scoping Results."

The staff sampled the contents of the UFSAR based on the systems and structures listed in LRA Table 2.2-1 to determine whether there were any systems or structures that may have intended functions within the scope of license renewal, as defined by 10 CFR 54.4, but had been omitted from the scope of license renewal. The staff identified no omissions.

2.2.4 Conclusion

The staff reviewed LRA Section 2.2 and the supporting information in the UFSAR to determine whether any systems and structures within the scope of license renewal had not been identified by the applicant. No omissions were identified. On the basis of this review, the staff concludes that there is reasonable assurance that the applicant has adequately identified the systems and structures within the scope of license renewal, in accordance with 10 CFR 54.4.

2.3 Scoping and Screening Results: Mechanical

This section documents the staff's review of the applicant's scoping and screening results for mechanical systems. Specifically, this section discusses the following systems:

- reactor vessel, internals, and reactor coolant system (RCS)
- engineered safety feature (ESF) systems
- auxiliary systems
- steam and power conversion systems

In accordance with the requirements of 10 CFR 54.21(a)(1), the applicant must identify and list passive, long-lived SCs within the scope of license renewal and subject to an AMR. To verify that the applicant properly implemented its methodology, the staff focused its review on the implementation results. This approach allowed the staff to confirm that there were no omissions of mechanical system components that meet the scoping criteria and are subject to an AMR.

Staff Evaluation Methodology. The staff's evaluation of the information in the LRA was the same for all mechanical systems. The objective was to determine whether the components and supporting structures for a specific system, that appeared to meet the scoping criteria specified in the Rule, had been identified by the applicant as within the scope of license renewal, in accordance with 10 CFR 54.4. Similarly, the staff evaluated the applicant's screening results to verify that all long-lived, passive components were subject to an AMR in accordance with 10 CFR 54.21(a)(1).

Scoping. To perform its evaluation, the staff reviewed the applicable LRA sections and associated component drawings, focusing on components that had not been identified as within the scope of license renewal. The staff reviewed relevant licensing basis documents, including the UFSAR, for each mechanical system to determine whether the applicant had omitted components with intended functions under 10 CFR 54.4(a) from the scope of license renewal. The staff also reviewed the licensing basis documents to determine whether all intended functions under 10 CFR 54.4(a) had been specified in the LRA. If omissions were identified, the staff requested additional information to resolve them.

Screening. After completing its review of the scoping results, the staff evaluated the applicant's screening results. For those SCs with intended functions, the staff sought to determine whether (1) the functions are performed with moving parts or a change in configuration or properties or (2) they are subject to replacement based on a qualified life or specified time period, as described in 10 CFR 54.21(a)(1). For those meeting neither of these criteria, the staff sought to confirm that these SCs were subject to an AMR, as required by 10 CFR 54.21(a)(1). If discrepancies were identified, the staff requested additional information to resolve them.

Two-Tier Scoping Review Process for Balance of Plant Systems. In the LRA there are 80 mechanical systems of which 31 are balance of plant (BOP) systems that include most of the auxiliary and all the steam and power conversion systems. The staff performed a two-tier scoping review for these BOP systems.

In the two-tier scoping review, the staff reviewed the LRA and UFSAR description focusing on the system intended function to screen all the BOP systems into two groups based on the following screening criteria:

- safety importance/risk significance
- potential for system failure to cause failure of redundant safety system trains
- operating experience indicating likely passive failures
- systems subject to omissions based on previous LRA reviews

Examples of safety and risk significant systems are the feedwater, the emergency diesel generator (EDG), auxiliary, and the emergency service water (ESW) systems based on the individual plant examination results for OCGS. An example of a system the failure of which could cause failure of redundant trains is a drain system for flood protection. Examples of systems with operating experience indicating likely passive failures include the main steam, feedwater, and ESW systems. Examples of systems with omissions identified in previous LRA reviews include spent fuel cooling system and makeup water sources to safety systems.

From the 31 BOP systems, the staff selected 16 systems for a Tier-2 (detailed) scoping review as described above. For the remaining 15 BOP systems, the staff performed a Tier-1 (not requiring detailed boundary drawings) review of the LRA and UFSAR that would identify apparently missing components for an AMR. However, Tier-2 requires the review of detailed boundary drawings in accordance with SRP-LR Section 2.3. The following is a list of the 15 Tier-1 systems:

- chlorination system
- condensate system
- cranes and hoists
- fuel storage and handling system
- heating and process steam system
- main condenser
- main fuel oil storage and transfer system
- main generator and auxiliary system
- main turbine and auxiliary systems
- miscellaneous floor and equipment drain system
- process sampling system
- radiation monitoring system
- reactor building floor and equipment drains
- roof drains and overboard discharge system
- sanitary waste system

The staff verified that there is no risk-significant system in this list by examining the results of the OCGS integrated plant assessment (IPA). None of the 15 systems is a dominant contributor to core damage frequency (CDF), nor are these systems involved in the dominant initiating events.

The following lists the 16 Tier-2 systems:

- circulating water system
- drywell floor and equipment drains
- emergency diesel generator and auxiliary system
- emergency service water system
- instrument (control) air system
- nitrogen supply system
- post-accident sampling system
- reactor building closed cooling water system
- reactor water cleanup system
- service water system
- spent fuel pool cooling system
- turbine building closed cooling water system
- water treatment and distribution system
- condensate transfer system
- feedwater system
- main steam system

2.3.1 Reactor Vessel, Internals, and Reactor Coolant System

In LRA Section 2.3.1, the applicant identified the SCs of the reactor vessel, internals, and RCS subject to an AMR for license renewal.

The applicant described the supporting SCs of the reactor vessel, internals, and RCS in the following sections of the LRA:

- 2.3.1.1   control rods
- 2.3.1.2   fuel assemblies
- 2.3.1.3   isolation condenser system
- 2.3.1.4   nuclear boiler instrumentation
- 2.3.1.5   reactor head cooling system
- 2.3.1.6   reactor internals
- 2.3.1.7   reactor pressure vessel
- 2.3.1.8   reactor recirculation system

The staff's review findings on LRA Sections 2.3.1.1 – 2.3.1.8 are presented in SER Sections 2.3.1.1 – 2.3.1.8, respectively.

2.3.1.1 Control Rods

2.3.1.1.1 Summary of Technical Information in the Application

In LRA Section 2.3.1.1, the applicant described the control rods. The control rods are replaceable, mechanical components consisting of cruciform-shaped stainless steel assemblies containing neutron-absorbing material, designed for flux shaping and for reactivity control during reactor startup, power level changes, and shutdown. The reactor contains 137 control rods the purpose of which is to absorb neutrons in the reactor core, thereby providing the means to adjust core power shape, compensate for reactivity changes caused by fuel and burnable poison depletion, and fully shut down the nuclear reaction. They accomplish this purpose, in conjunction with their positioning system (evaluated with the control rod drive system), by providing continuous regulation of the core excess reactivity and reactivity distribution and by providing sufficient reactivity compensation to render the reactor adequately subcritical from its most reactive condition. Control rod absorption of neutrons chemically depletes the absorber material and control rod lifetime is monitored. Control rods reaching prescribed thresholds are scheduled for replacement during refueling outages.

The control rods contain safety-related components relied upon to remain functional during and following DBEs.

No intended functions within the scope of license renewal are applicable for the controls rods.

In LRA Table 2.3.1.1, the applicant identified no control rods component types within the scope of license renewal and subject to an AMR because all components are short-lived.

2.3.1.1.2 Staff Evaluation

The staff reviewed LRA Section 2.3.1.1 and UFSAR Sections 4.3.2.4 and 4.6.4.3 using the evaluation methodology of SER Section 2.3. The staff conducted its review in accordance with the guidance of SRP-LR Section 2.3, "Scoping and Screening Results: Mechanical Systems."

In conducting its review, the staff evaluated the system functions described in the LRA and UFSAR to verify that the applicant had not omitted from the scope of license renewal any components with intended functions under 10 CFR 54.4(a). The staff then reviewed those components that the applicant had identified as within the scope of license renewal to verify that it had not omitted any passive and long-lived components subject to an AMR in accordance with the requirements of 10 CFR 54.21(a)(1).

2.3.1.1.3 Conclusion

The staff reviewed the LRA to determine whether any SSCs that should be within the scope of license renewal had not been identified by the applicant. No omissions were identified. In addition, the staff determined whether any components subject to an AMR had not been identified by the applicant. No omissions were identified. The staff concludes that there is reasonable assurance that the applicant has adequately identified the control rods components within the scope of license renewal, as required by 10 CFR 54.4(a), and those subject to an AMR, as required by 10 CFR 54.21(a)(1).

2.3.1.2 Fuel Assemblies

2.3.1.2.1 Summary of Technical Information in the Application

In LRA Section 2.3.1.2, the applicant described the fuel assemblies, high-integrity components containing the fissionable material that sustains the nuclear reaction when the reactor core is made critical. The purpose of the fuel assemblies is to allow efficient heat transfer from the nuclear fuel to the reactor coolant and to maintain structural integrity providing a controllable, coolable bundle geometry and fission product barrier. They accomplish this purpose by satisfying the thermal-mechanical, nuclear, and hydraulic requirements of the nuclear fuel design conditions within the reactor. Each fuel assembly is comprised of a fuel bundle and a channel that surrounds it. The fuel rods of each bundle are spaced and supported in a square array by the stainless steel upper and lower tie plates and intermediately placed zircaloy spacer assemblies. The bundle channel is fabricated from zircaloy and provides the flow path outer periphery for bundle coolant flow, supplies structural stiffness to the bundle and transmits seismic loadings to the core internal structures, provides a heat sink during a loss of cooling accident (LOCA), and supplies a surface for control rod guidance within the reactor core. The reactor contains 560 fuel bundle assemblies. During each refueling outage, approximately one-third of the highest depletion bundles are replaced and the positions of the remaining bundles are shuffled as required by the nuclear core design to optimize cycle energy, operating conditions, and fuel economics. Cycle-specific evaluations of the thermal mechanical design known as supplemental reload licensing submittals are produced to ensure that the safety and operational requirements of the fuel product line are met.

The fuel assemblies contain safety-related components relied upon to remain functional during and following DBEs.

No intended functions within the scope of license renewal are applicable for the fuel assemblies.

In LRA Table 2.3.1.2, the applicant identified no fuel assembly component types within the scope of license renewal and subject to an AMR because all components are short-lived.

2.3.1.2.2 Staff Evaluation

The staff reviewed LRA Section 2.3.1.2 and UFSAR Section 4.2.2 using the evaluation methodology of SER Section 2.3. The staff conducted its review in accordance with the guidance of SRP-LR Section 2.3.

In conducting its review, the staff evaluated the system functions described in the LRA and UFSAR to verify that the applicant had not omitted from the scope of license renewal any components with intended functions under 10 CFR 54.4(a). The staff then reviewed those components that the applicant had identified as within the scope of license renewal to verify that it had not omitted any passive and long-lived components subject to an AMR in accordance with the requirements of 10 CFR 54.21(a)(1).

2.3.1.2.3 Conclusion

The staff reviewed the LRA to determine whether any SSCs that should be within the scope of license renewal had not been identified by the applicant. No omissions were identified. In addition, the staff determined whether any components subject to an AMR had not been identified by the applicant. No omissions were identified. The staff concludes that there is reasonable assurance that the applicant has adequately identified the fuel assemblies components within the scope of license renewal, as required by 10 CFR 54.4(a), and those subject to an AMR, as required by 10 CFR 54.21(a)(1).

2.3.1.3 Isolation Condenser System

2.3.1.3.1 Summary of Technical Information in the Application

In LRA Section 2.3.1.3, the applicant described the ICS. The ICS is a standby, high-pressure system designed for removal of fission product decay heat when the reactor vessel is isolated from the main condenser. This condition can occur when the main steam isolation valves (MSIVs) have closed or the main condenser is otherwise unavailable for use as a heat sink. The purpose of the system is to prevent overheating of the reactor fuel, control the reactor pressure rise, and limit the loss of reactor coolant through the relief valves. The ICS accomplishes this purpose by depressurizing the reactor and removing residual and decay heat. ICS operation is initiated automatically by reactor vessel high pressure or low-low water level or can be initiated manually. The ICS is comprised of two independent loops, each with one condenser shell containing two tube bundles. When a loop is in operation, both tube bundles are in service. For ICS initiation, normally both condensers are placed in operation simultaneously, and either loop can be activated or shut down separately by manual control. The ICS operates by natural circulation without the need for driving power other than the direct current (DC) electrical system used to open an isolation valve on each condensate return line, initiating ICS operation.

The ICS contains safety-related components relied upon to remain functional during and following DBEs. The failure of nonsafety-related SSCs in the ICS could potentially prevent the satisfactory accomplishment of a safety-related function. In addition, the ICS performs functions that support fire protection, SBO and EQ.

The intended functions within the scope of license renewal include:

- provides filtration

- provides heat transfer

- maintains mechanical and structural integrity to prevent spatial interactions that could cause failure of safety-related SSCs (includes the required structural support when the nonsafety-related leakage boundary piping is also attached to safety-related piping)

- provides mechanical closure

- provides pressure-retaining boundary; fission product barrier; containment isolation; or containment, holdup, and plateout (main steam system)

In LRA Table 2.3.1.3, the applicant identified the following ICS component types within the scope of license renewal and subject to an AMR:

- bird screen
- closure bolting
- gauge snubber
- heat exchangers (isolation condensers)
- piping and fittings
- thermowell
- valve body

2.3.1.3.2 Staff Evaluation

The staff reviewed LRA Section 2.3.1.3 and UFSAR Sections 3.6.2.6 and 6.3 using the evaluation methodology of SER Section 2.3. The staff conducted its review in accordance with the guidance of SRP-LR Section 2.3.

In conducting its review, the staff evaluated the system functions described in the LRA and UFSAR to verify that the applicant had not omitted from the scope of license renewal any components with intended functions under 10 CFR 54.4(a). The staff then reviewed those components that the applicant had identified as within the scope of license renewal to verify that it had not omitted any passive and long-lived components subject to an AMR in accordance with the requirements of 10 CFR 54.21(a)(1).

2.3.1.3.3 Conclusion

The staff reviewed the LRA to determine whether any SSCs that should be within the scope of license renewal had not been identified by the applicant. No omissions were identified. In addition, the staff determined whether any components subject to an AMR had not been identified by the applicant. No omissions were identified. The staff concludes that there is reasonable assurance that the applicant has adequately identified the ICS components within the scope of license renewal, as required by 10 CFR 54.4(a), and those subject to an AMR, as required by 10 CFR 54.21(a)(1).

2.3.1.4 Nuclear Boiler Instrumentation

2.3.1.4.1 Summary of Technical Information in the Application

In LRA Section 2.3.1.4, the applicant described the nuclear boiler instrumentation. The nuclear boiler instrumentation system is designed to provide the means to measure parameters of level, pressure, temperature, flow, core differential pressure, and core spray pipe integrity. The purpose of the system is to provide signals to the reactor protection system and emergency core cooling system (ECCS) logic for initiation of such protective system functions as reactor scram, ECCS and Engineered Safety Feature (ESF) system initiation, primary containment isolation, recirculation pump trip, and alternate rod insertion. The feedwater control function is provided input from this system. Nuclear boiler instrumentation also provides the operator with indications of reactor level, pressure, temperature, and flow during normal and transient conditions to support procedural activities during normal and post-accident operation. It accomplishes these purposes by utilizing specific instruments to monitor level, pressure (including differential pressure), flow, and temperature. Reactor vessel level is measured by comparing the differential pressure between the variable level of water in the reactor vessel and the pressure from a reference water column of a known height. Reactor pressure is measured by pressure instruments utilizing the same piping used to measure the pressure in the water level instrument reference legs. Temperature is measured through thermocouples placed in specific locations on the reactor vessel shell, heads, flange, and skirt to indicate vessel metal temperature.

The nuclear boiler instrumentation contains safety-related components relied upon to remain functional during and following DBEs. The failure of nonsafety-related SSCs in the nuclear boiler instrumentation potentially could prevent the satisfactory accomplishment of a safety-related function. In addition, the nuclear boiler instrumentation performs functions for fire protection, ATWS, SBO, and EQ.

The intended functions within the scope of license renewal include:

- maintains mechanical and structural integrity to prevent spatial interactions that could cause failure of safety-related SSCs (includes the required structural support when the nonsafety-related leakage boundary piping is also attached to safety-related piping)
- provides mechanical closure
- provides pressure-retaining boundary; fission product barrier; containment isolation; or containment, holdup, and plateout (main steam system)

In LRA Table 2.3.1.4, the applicant identified the following nuclear boiler instrumentation component types within the scope of license renewal and subject to an AMR:

- closure bolting
- condensing chamber
- gauge snubber
- piping and fittings
- valve body

2.3.1.4.2 Staff Evaluation

The staff reviewed LRA Section 2.3.1.4 and UFSAR Section 7.6.1.1 using the evaluation methodology of SER Section 2.3. The staff conducted its review in accordance with the guidance of SRP-LR Section 2.3.

In conducting its review, the staff evaluated the system functions described in the LRA and UFSAR to verify that the applicant had not omitted from the scope of license renewal any components with intended functions under 10 CFR 54.4(a). The staff then reviewed those components that the applicant had identified as within the scope of license renewal to verify that it had not omitted any passive and long-lived components subject to an AMR in accordance with the requirements of 10 CFR 54.21(a)(1).

2.3.1.4.3 Conclusion

The staff reviewed the LRA to determine whether any SSCs that should be within the scope of license renewal had not been identified by the applicant. No omissions were identified. In addition, the staff determined whether any components subject to an AMR had not been identified by the applicant. No omissions were identified. The staff concludes that there is reasonable assurance that the applicant has adequately identified the nuclear boiler instrumentation components within the scope of license renewal, as required by 10 CFR 54.4(a), and those subject to an AMR, as required by 10 CFR 54.21(a)(1).

2.3.1.5 Reactor Head Cooling System

2.3.1.5.1 Summary of Technical Information in the Application

In LRA Section 2.3.1.5, the applicant described the reactor head cooling system (RHCS) designed for use in conjunction with reactor vessel flooding and the shutdown cooling system (SCS) for condensing steam formed in the vessel head and for cooling the flanges and the upper portions of the reactor pressure vessel during shutdown operation. The RHCS condenses

2-38

steam and condensable gases in the vessel dome to assist in vessel head cooling during shutdown, prevents repressurization as the vessel is flooded to levels above the vessel flange and main steam nozzles to cool the upper portions of the vessel metal, and permits reactor pressure to be reduced to atmospheric while reducing vessel head temperature. A cross-connect line between the head cooling line and the head vent line prevents accumulation of hydrogen and other non-condensable gases in the head cooling line above the reactor vessel during normal power operation. The RHCS is comprised of a single spray nozzle located inside the top of the reactor pressure vessel head spraying through a cone angle which does not strike the head metal surface. The head spray water is supplied by the standby control rod drive (CRD) system feed pump.

The RHCS contains safety-related components relied upon to remain functional during and following DBEs. The failure of nonsafety-related SSCs in the RHCS potentially could prevent the satisfactory accomplishment of a safety-related function. In addition, the RHCS performs functions that support EQ.

The intended functions within the scope of license renewal include:

- maintains mechanical and structural integrity to prevent spatial interactions that could cause failure of safety-related SSCs (includes the required structural support when the nonsafety-related leakage boundary piping is also attached to safety-related piping)

- provides mechanical closure

- provides pressure-retaining boundary; fission product barrier; containment isolation; or containment, holdup, and plateout (main steam system)

- provides flow restriction

In LRA Table 2.3.1.5, the applicant identified the following RHCS component types within the scope of license renewal and subject to an AMR:

- closure bolting
- flow element
- piping and fittings
- restricting orifice
- valve body

2.3.1.5.2 Staff Evaluation

The staff reviewed LRA Section 2.3.1.5 and UFSAR Section 5.4.11 using the evaluation methodology of SER Section 2.3. The staff conducted its review in accordance with the guidance of SRP-LR Section 2.3.

In conducting its review, the staff evaluated the system functions described in the LRA and UFSAR to verify that the applicant had not omitted from the scope of license renewal any components with intended functions under 10 CFR 54.4(a). The staff then reviewed those components that the applicant had identified as within the scope of license renewal to verify that it had not omitted any passive and long-lived components subject to an AMR in accordance with the requirements of 10 CFR 54.21(a)(1).

2.3.1.5.3 Conclusion

The staff reviewed the LRA to determine whether any SSCs that should be within the scope of license renewal had not been identified by the applicant. No omissions were identified. In addition, the staff determined whether any components subject to an AMR had not been identified by the applicant. No omissions were identified. The staff concludes that there is reasonable assurance that the applicant has adequately identified the RHCS components within the scope of license renewal, as required by 10 CFR 54.4(a), and those subject to an AMR, as required by 10 CFR 54.21(a)(1).

2.3.1.6 Reactor Internals

2.3.1.6.1 Summary of Technical Information in the Application

In LRA Section 2.3.1.6, the applicant described the reactor internals. The reactor internals support the core and other internal components, maintain the fuel in a coolable geometry during normal and accident conditions, and properly distribute the coolant delivered to the vessel. Major components of the reactor internals include the shroud, steam separator assembly, recirculation outlet, inlet plenum, shroud support ring, cone support ring, upper core grid (top guide), bottom core support plate, and the peripheral fuel assemblies. The shroud is a stainless steel cylinder that surrounds the reactor core and provides a barrier to separate the upward flow of the coolant through the reactor core from the downward recirculation flow. Bolted on top of the shroud is the steam separator assembly, which forms the top of the core discharge plenum and provides a mixing chamber for the steam-water mixture before it enters the steam separator. The recirculation outlet and inlet plenum are separated by the shroud support ring (support cone), which joins the bottom of the shroud to the vessel wall. The cone support ring carries all the vertical weight of the shroud, steam separator and dryer assembly, upper core grid (top guide), bottom core support plate, and the peripheral fuel assemblies. The shroud support ring also sustains the differential upward pressure loading on the shroud under operating conditions and the vertical and lateral seismic loads developed during an earthquake. The control rod guide tubes extend up from the control rod drive housing through holes in the core plate. Each tube is designed as a lateral guide for the control rod and as the vertical support for the fuel support piece, which holds the four fuel assemblies surrounding the control rod. Except for the weight of the peripheral fuel assemblies, the entire weight of the fuel is carried by the guide tubes and transmitted to the bottom head through the CRD housings and stub tubes.

The reactor internals contain safety-related components relied upon to remain functional during and following DBEs. The failure of nonsafety-related SSCs in the reactor internals potentially could prevent the satisfactory accomplishment of a safety-related function. In addition, the reactor internals performs functions that support fire protection.

The intended functions within the scope of license renewal include:

- provides spray shield or curbs for directing flow
- provides pressure-retaining boundary; fission product barrier; containment isolation; or containment, holdup, and plateout (main steam system)
- provides conversion of fluid into spray

2-40

- provides structural support or structural integrity to preclude nonsafety-related component interactions that could prevent satisfactory accomplishment of a safety-related function

In LRA Table 2.3.1.6, the applicant identified the following reactor internals component types within the scope of license renewal and subject to an AMR:

- CRD assembly (housing and guide tube)
- core plate (lower core grid)
- core plate (lower core grid) wedges
- core shroud
- core spray line spray nozzle elbows
- core spray lines, thermal sleeves, spray rings (sparger), and spray nozzles
- core spray ring (sparger) repair hardware
- fuel support piece
- incore neutron monitor dry tubes, guide tubes, and housings
- shroud repairs (tie rods and lug/clevis assemblies)
- shroud support structure
- top guide (upper core grid)
- vessel steam dryer

2.3.1.6.2 Staff Evaluation

The staff reviewed LRA Section 2.3.1.6 and UFSAR Sections 3.9.5 and 4.5.2 using the evaluation methodology of SER Section 2.3. The staff conducted its review in accordance with the guidance of SRP-LR Section 2.3.

In conducting its review, the staff evaluated the system functions described in the LRA and UFSAR to verify that the applicant had not omitted from the scope of license renewal any components with intended functions under 10 CFR 54.4(a). The staff then reviewed those components that the applicant had identified as within the scope of license renewal to verify that it had not omitted any passive and long-lived components subject to an AMR in accordance with the requirements of 10 CFR 54.21(a)(1).

The staff's review of the LRA identified an area in which additional information was necessary to complete the review of the applicant's scoping and screening results. The applicant responded to the staff's RAI as discussed below.

In RAI 2.3.1.6-1 dated March 10, 2006, the staff noted that LRA Section 2.3.1.6 states that the reactor vessel head spray nozzle supports no intended functions delineated in the Rule and, therefore, is not included within the scope of license renewal, and that a safety assessment for this component was performed and reported in Boiling Water Reactor Vessel and Internals Project (BWRVIP)-06. However, the staff could not locate the safety assessment in the referenced document. Therefore, the staff requested clarification.

In its response dated April 7, 2006, the applicant agreed that the BWRVIP-06 does not include an assessment of the reactor vessel head spray nozzle as stated in LRA Section 2.3.1.6; therefore, the reference to BWRVIP-06 was an error. The applicant, however, added that the head spray nozzle performs no safety-related function, that it is not credited for any regulated event, and that no postulated failure of the head spray nozzle could cause failure of safety-related equipment. Therefore, the applicant maintains its position as stated in the LRA, that the head spray nozzle supports no intended functions and is not included within the scope of license renewal.

During a teleconference April 7, 2006, the information provided in the UFSAR supplement on the reactor pressure vessel (RPV) head cooling system was discussed. The applicant stated that the nozzle does not meet the criteria for in-scope components and is used only for normal shutdown. The applicant and the staff also discussed the requirements identified in the UFSAR and in 10 CFR Part 50, Appendix R. The applicant stated that the nozzle is not needed to meet Appendix R safe shutdown requirements. The staff understood the applicant to exclude the nozzle from the scope of license renewal and concludes that the response was acceptable. The staff's concern described in RAI 2.3.1.6-1 is resolved.

2.3.1.6.3 Conclusion

The staff reviewed the LRA and the RAI response to determine whether any SSCs that should be within the scope of license renewal had not been identified by the applicant. No omissions were identified. In addition, the staff determined whether any components subject to an AMR had not been identified by the applicant. No omissions were identified. The staff concludes that there is reasonable assurance that the applicant has adequately identified the reactor internals components within the scope of license renewal, as required by 10 CFR 54.4(a), and those subject to an AMR, as required by 10 CFR 54.21(a)(1).

2.3.1.7 Reactor Pressure Vessel

2.3.1.7.1 Summary of Technical Information in the Application

In LRA Section 2.3.1.7, the applicant described the RPV, which contains the reactor core, the reactor internals, and reactor core coolant moderator. The RPV forms part of the reactor coolant pressure boundary (RCPB) and serves as a high-integrity barrier against leakage of radioactive materials to the drywell. The RPV is a vertical, cylindrical pressure vessel with hemispherical heads. The cylindrical shell and bottom hemispherical head of the RPV are of welded construction fabricated of low-alloy steel plate. The removable top head attached to the cylindrical shell flange with studs and nuts includes two concentric seal rings in the head flange. The RPV is supported by a steel skirt, the top of which is welded to the bottom of the vessel. The base of the skirt is continuously supported by a ring girder fastened to a concrete foundation, which carries the load through the drywell to the reactor building foundation slab. The major RPV safety function is to provide a radioactive material barrier.

The RPV contains safety-related components relied upon to remain functional during and following DBEs. In addition, the RPV performs functions that support fire protection.

The intended functions within the scope of license renewal include:

- provides spray shield or curbs for directing flow

- provides mechanical closure

- provides pressure-retaining boundary; fission product barrier; containment isolation; or containment, holdup, and plateout (main steam system)

- provides structural support or structural integrity to preclude nonsafety-related component interactions that could prevent satisfactory accomplishment of a safety-related function

In LRA Table 2.3.1.7, the applicant identified the following RPV component types within the scope of license renewal and subject to an AMR:

- nozzle (bottom head drain)
- nozzle safe ends (core spray, isolation condenser, and CRD return)
- nozzle safe ends (feedwater and main steam)
- nozzle safe ends (recirculation inlet and outlet)
- nozzle thermal sleeves (CRD return line)
- nozzle thermal sleeves (feedwater nozzle)
- nozzles (core spray)
- nozzles (CRD return)
- nozzles (feedwater)
- nozzles (main steam and isolation condenser)
- nozzles (recirculation inlet and outlet)
- penetrations (CRD stub tubes)
- penetrations (instrumentation including safe ends)
- penetrations (standby liquid control)
- RPV support skirt and attachment welds
- top head closure studs and nuts
- top head enclosure (head and nozzles)
- top head enclosure vessel flange leak detection penetration
- top head flange
- vessel bottom head
- vessel shell (upper, upper intermediate, lower intermediate, lower, and belt line welds)
- vessel shell attachment welds
- vessel shell flange

2.3.1.7.2 Staff Evaluation

The staff reviewed LRA Section 2.3.1.7 and UFSAR Sections 5.1 and 5.3 using the evaluation methodology of SER Section 2.3. The staff conducted its review in accordance with the guidance of SRP-LR Section 2.3.

In conducting its review, the staff evaluated the system functions described in the LRA and UFSAR to verify that the applicant had not omitted from the scope of license renewal any components with intended functions under 10 CFR 54.4(a). The staff then reviewed those components that the applicant had identified as within the scope of license renewal to verify that it had not omitted any passive and long-lived components subject to an AMR in accordance with the requirements of 10 CFR 54.21(a)(1).

The staff's review of the LRA identified an area in which additional information was necessary to complete the review of the applicant's scoping and screening results. The applicant responded to the staff's RAI as discussed below.

In RAI 2.3.1.7-1 dated March 10, 2006, the staff noted that LRA Table 2.3.1.7 lists the component type "Top Head Enclosure Vessel Flange Leak Detection Penetration" as within the scope of license renewal and subject to an AMR. However, it was not clear whether the tubes/pipes connected to the penetration also were included within the scope of license renewal. Therefore, the staff requested that the applicant confirm whether the subject tubes/pipes had been included within the scope of license renewal and, if not, that the applicant include the subject components within the scope of license renewal requiring an AMR.

In its response dated April 7, 2006, the applicant stated that the vessel leak-off piping was included within the scope of license renewal and considered part of the nuclear boiler instrumentation system. The applicant further stated that the subject component was in LRA Table 2.3.1.4 in component types "pipings and fittings" and "valve body." The staff finds the response acceptable. The staff's concern described in RAI 2.3.1.7-1 is resolved.

2.3.1.7.3 Conclusion

The staff reviewed the LRA and the RAI response to determine whether any SSCs that should be within the scope of license renewal had not been identified by the applicant. No omissions were identified. In addition, the staff determined whether any components subject to an AMR had not been identified by the applicant. No omissions were identified. The staff concludes that there is reasonable assurance that the applicant has adequately identified the RPV components within the scope of license renewal, as required by 10 CFR 54.4(a), and those subject to an AMR, as required by 10 CFR 54.21(a)(1).

2.3.1.8 Reactor Recirculation System

2.3.1.8.1 Summary of Technical Information in the Application

In LRA Section 2.3.1.8, the applicant described the reactor recirculation system, a reactivity control system that provides forced circulation of reactor coolant through the core. The reactor recirculation system consists of the reactor recirculation main loop piping, recirculation pumps and motors, recirculation motor-generator sets, recirculation system flow control, and recirculation pump trip logic. The purpose of the reactor recirculation system, to provide forced circulation of reactor coolant through the core, controls reactor power within a limited range without the need for manipulation of the control rods. It accomplishes this purpose by delivering recirculated water flow to the reactor vessel through five separate pumped loops, each with an individually controllable variable speed pump. Under normal reactor power conditions, all five recirculation loops are in operation, with all pumps operating at the same speed. Plant operation has been analyzed with up to two recirculation loops out of service. Recirculation pump trip (RPT) is an instrument-controlled function of the reactor recirculation system that decreases the pressure and temperature transient during an ATWS event. The reactor protection system (RPS) supplies a signal to the RPT system causing a trip of all five recirculation pumps on a vessel low-low level signal. On a vessel high-pressure signal from RPS, RPT trips three recirculation pumps immediately and trips the remaining two pumps after a timed delay if the vessel high-pressure condition still exists.

The reactor recirculation system contains safety-related components relied upon to remain functional during and following DBEs. The failure of nonsafety-related SSCs in the reactor recirculation system potentially could prevent the satisfactory accomplishment of a safety-related function. In addition, the reactor recirculation system performs functions that support fire protection, ATWS, and SBO.

2-44

The intended functions within the scope of license renewal include:

- maintains mechanical and structural integrity to prevent spatial interactions that could cause failure of safety-related SSCs (includes the required structural support when the nonsafety-related leakage boundary piping is also attached to safety-related piping)
- provides mechanical closure
- provides pressure-retaining boundary; fission product barrier; containment isolation; or containment, holdup, and plateout (main steam system)

In LRA Table 2.3.1.8, the applicant identified the following reactor recirculation system component types within the scope of license renewal and subject to an AMR:

- closure bolting
- coolers (oil)
- filter housing (oil)
- flow element
- fluid drive (MG set coupling) - reservoir
- oil mist eliminator - reservoir
- piping and fittings
- pump casing
- sight glasses (oil)
- thermowell
- valve body

2.3.1.8.2 Staff Evaluation

The staff reviewed LRA Section 2.3.1.8 and UFSAR Sections 5.4.1, 7.6.1, and 7.6.2 using the evaluation methodology of SER Section 2.3. The staff conducted its review in accordance with the guidance of SRP-LR Section 2.3.

In conducting its review, the staff evaluated the system functions described in the LRA and UFSAR to verify that the applicant had not omitted from the scope of license renewal any components with intended functions under 10 CFR 54.4(a). The staff then reviewed those components that the applicant had identified as within the scope of license renewal to verify that it had not omitted any passive and long-lived components subject to an AMR in accordance with the requirements of 10 CFR 54.21(a)(1).

2.3.1.8.3 Conclusion

The staff reviewed the LRA to determine whether any SSCs that should be within the scope of license renewal had not been identified by the applicant. No omissions were identified. In addition, the staff determined whether any components subject to an AMR had not been identified by the applicant. No omissions were identified. The staff concludes that there is reasonable assurance that the applicant has adequately identified the reactor recirculation system components within the scope of license renewal, as required by 10 CFR 54.4(a), and those subject to an AMR, as required by 10 CFR 54.21(a)(1).

2.3.2 Engineered Safety Features Systems

In LRA Section 2.3.2, the applicant identified the SCs of the ESF systems subject to an AMR for license renewal.

The applicant described the supporting SCs of the ESF systems in the following sections of the LRA:

- 2.3.2.1   automatic depressurization system
- 2.3.2.2   containment spray system
- 2.3.2.3   core spray system
- 2.3.2.4   standby gas treatment system (SGTS)

The staff findings on LRA Sections 2.3.2.1 – 2.3.2.4 are presented in SER Sections 2.3.2.1 – 2.3.2.4, respectively.

2.3.2.1 Automatic Depressurization System

2.3.2.1.1 Summary of Technical Information in the Application

In LRA Section 2.3.2.1, the applicant described the automatic depressurization system (ADS), a standby ECCS designed to provide a controlled blowdown of the primary system to rapidly reduce pressure during a small pipe break. Depressurization following a LOCA permits the low-pressure core spray system to achieve timely rated flow of injection water into the reactor core to prevent fuel clad melting. For larger breaks the vessel depressurizes sufficiently to permit core spray injection without ADS assistance. The ADS equipment also provides an overpressure protection function for the RPV. The ADS is one of the systems that comprise the ECCS and as such is designed to function throughout the post-accident period. The purpose of the ADS is to depressurize the RCS either during a small break LOCA or in the event of an overpressure condition in the RPV. The ADS accomplishes this purpose by opening the electromatic relief valves (EMRVs) to provide a controlled blowdown of the primary coolant system and rapidly reduce reactor vessel pressure during a small pipe break or overpressure condition. Additionally, manual ADS actuation of the EMRVs is credited for pressure control during an isolation condenser high-energy line break. The ADS automatic depressurization function, the overpressure function, and the manual operation of the EMRVs are all controlled through the ADS logic network.

The ADS contains safety-related components relied upon to remain functional during and following DBEs. In addition, the ADS performs functions that support fire protection, SBO, and EQ.

The intended functions within the scope of license renewal include:

- provides emergency core cooling where the equipment provides coolant directly to the core

- provides an RCPB

- provides a sensor of process conditions and generates signals for reactor trip or ESF actuation

- relied upon in safety analyses or plant evaluations to perform a function that demonstrates compliance with the Commission's regulations for fire protection (10 CFR 50.48)

- relied upon in safety analyses or plant evaluations to perform a function that demonstrates compliance with the Commission's regulations for EQ (10 CFR 50.49)

- relied upon in safety analyses or plant evaluations to perform a function that demonstrates compliance with the Commission's regulations for SBO (10 CFR 50.63)

The applicant identified the following ADS component types, which are evaluated with the main steam system (LRA Section 2.3.4.6), within the scope of license renewal and subject to an AMR:

- EMRV assemblies
- vacuum breakers
- piping and associated components
- Y-quenchers located in the torus

2.3.2.1.2 Staff Evaluation

The staff reviewed LRA Section 2.3.2.1 and UFSAR Sections 3.6.2.6.1, 5.2.2, 6.3.1.2, and 7.3.1 using the evaluation methodology of SER Section 2.3. The staff conducted its review in accordance with the guidance of SRP-LR Section 2.3.

In conducting its review, the staff evaluated the system functions described in the LRA and UFSAR to verify that the applicant had not omitted from the scope of license renewal any components with intended functions under 10 CFR 54.4(a). The staff then reviewed those components that the applicant had identified as within the scope of license renewal to verify that it had not omitted any passive and long-lived components subject to an AMR in accordance with the requirements of 10 CFR 54.21(a)(1).

2.3.2.1.3 Conclusion

The staff reviewed the LRA to determine whether any SSCs that should be within the scope of license renewal had not been identified by the applicant. No omissions were identified. In addition, the staff determined whether any components subject to an AMR had not been identified by the applicant. No omissions were identified. The staff concludes that there is reasonable assurance that the applicant has adequately identified the ADS components within the scope of license renewal, as required by 10 CFR 54.4(a), and those subject to an AMR, as required by 10 CFR 54.21(a)(1).

2.3.2.2 Containment Spray System

2.3.2.2.1 Summary of Technical Information in the Application

In LRA Section 2.3.2.2, the applicant described the containment spray system, a standby system designed to be used with the core spray system to remove the reactor core decay heat from the containment in the event of a LOCA. The ESW system cools the containment spray heat exchangers, thereby providing the heat sink for the energy released during a LOCA. The containment spray system has the alternate capability of cooling the water in the torus pool during normal, shutdown, and post-accident conditions. The containment spray system is comprised of two redundant loops that deliver water from the torus pool to the spray headers in the drywell and torus. The containment spray system is manually initiated from switches in the control room. The containment spray pumps can be started manually for containment spray service if the proper containment spray initiation permissives are met. Two independent mode select switches are provided, one for each loop, each with two modes, "drywell spray" and "torus cooling."

The containment spray system contains safety-related components relied upon to remain functional during and following DBEs. In addition, the containment spray system performs functions that support fire protection and EQ.

The intended functions within the scope of license renewal include:

- provides filtration

- maintains mechanical and structural integrity to prevent spatial interactions that could cause failure of safety-related SSCs (includes the required structural support when the nonsafety-related leakage boundary piping is also attached to safety-related piping)

- provides mechanical closure

- provides pressure-retaining boundary; fission product barrier; containment isolation; or containment, holdup, and plateout (main steam system)

- provides conversion of fluid into spray

In LRA Table 2.3.2.2, the applicant identified the following containment spray system component types within the scope of license renewal and subject to an AMR:

- closure bolting
- flow element
- piping and fittings
- pump casing
- spray nozzle
- strainer (ECCS suction)
- thermowell
- valve body

2.3.2.2.2 Staff Evaluation

The staff reviewed LRA Section 2.3.2.2 and UFSAR Sections 6.2.2 and 6.5.2 using the evaluation methodology of SER Section 2.3. The staff conducted its review in accordance with the guidance of SRP-LR Section 2.3.

In conducting its review, the staff evaluated the system functions described in the LRA and UFSAR to verify that the applicant had not omitted from the scope of license renewal any components with intended functions under 10 CFR 54.4(a). The staff then reviewed components that the applicant had identified as within the scope of license renewal to verify that it had not omitted any passive and long-lived components subject to an AMR in accordance with the requirements of 10 CFR 54.21(a)(1).

2.3.2.2.3 Conclusion

The staff reviewed the LRA to determine whether any SSCs that should be within the scope of license renewal had not been identified by the applicant. No omissions were identified. In addition, the staff determined whether any components subject to an AMR had not been identified by the applicant. No omissions were identified. The staff concludes that there is reasonable assurance that the applicant has adequately identified the containment spray system components within the scope of license renewal, as required by 10 CFR 54.4(a), and those subject to an AMR, as required by 10 CFR 54.21(a)(1).

2.3.2.3 Core Spray System

2.3.2.3.1 Summary of Technical Information in the Application

In LRA Section 2.3.2.3, the applicant described the core spray system, a low-pressure ECCS designed to provide cooling water for removal of decay heat from the reactor core following a LOCA. Large-to-intermediate pipe breaks in the RCS cause a reactor pressure reduction sufficient to permit the core spray system to achieve its rated injection flow prior to fuel cladding melt. To accommodate the remaining intermediate-to-small pipe breaks, the ADS provides the initial controlled depressurization to reduce reactor pressure and thus permit timely core spray injection. In this manner, the core spray system provides core cooling to prevent fuel clad melting for the entire spectrum of postulated LOCAs. The core spray system provides a supply of cooling water to the reactor core independent of the feedwater system and operable on emergency power. The core spray system is comprised of two independent loops, each containing full flow test, keep-fill, and minimum flow pump protection features. Initiation of both loops of the core spray system occurs upon receipt of a high drywell pressure or low-low reactor vessel level signal. These signals also start both EDGs to supply power to the core spray pumps in the event of loss of normal electric power supply. The core spray system also can be initiated manually.

The core spray system contains safety-related components relied upon to remain functional during and following DBEs. The failure of nonsafety-related SSCs in the core spray system potentially could prevent the satisfactory accomplishment of a safety-related function. In addition, the core spray system performs functions that support fire protection, SBO, and EQ.

The intended functions within the scope of license renewal include:

- maintains mechanical and structural integrity to prevent spatial interactions that could cause failure of safety-related SSCs (includes the required structural support when the nonsafety-related leakage boundary piping is also attached to safety-related piping)
- provides mechanical closure
- provides pressure-retaining boundary; fission product barrier; containment isolation; or containment, holdup, and plateout (main steam system)
- provides flow restriction

In LRA Table 2.3.2.3, the applicant identified the following core spray system component types within the scope of license renewal and subject to an AMR:

- closure bolting
- cyclone separator
- flow element
- gauge snubber
- piping and fittings
- pump casing (fill pumps)
- pump casing (main and booster pumps)
- restricting orifice
- sight glasses
- thermowell
- valve body

2.3.2.3.2 Staff Evaluation

The staff reviewed LRA Section 2.3.2.3 and UFSAR Sections 6.3.1 and 6.3.1.3 using the evaluation methodology of SER Section 2.3. The staff conducted its review in accordance with the guidance of SRP-LR Section 2.3.

In conducting its review, the staff evaluated the system functions described in the LRA and UFSAR to verify that the applicant had not omitted from the scope of license renewal any components with intended functions under 10 CFR 54.4(a). The staff then reviewed those components that the applicant had identified as within the scope of license renewal to verify that it had not omitted any passive and long-lived components subject to an AMR in accordance with the requirements of 10 CFR 54.21(a)(1).

2.3.2.3.3 Conclusion

The staff reviewed the LRA to determine whether any SSCs that should be within the scope of license renewal had not been identified by the applicant. No omissions were identified. In addition, the staff determined whether any components subject to an AMR had not been identified by the applicant. No omissions were identified. The staff concludes that there is reasonable assurance that the applicant has adequately identified the core spray system components within the scope of license renewal, as required by 10 CFR 54.4(a), and those subject to an AMR, as required by 10 CFR 54.21(a)(1).

2.3.2.4 Standby Gas Treatment System (SGTS)

2.3.2.4.1 Summary of Technical Information in the Application

In LRA Section 2.3.2.4, the applicant described the SGTS, a plant ESF ventilation system that filters and exhausts the reactor building atmosphere and drywell atmosphere to the stack during secondary containment isolation conditions and drywell purging operations. The purpose of the system is to limit post-accident radioactive releases to the environs by collecting, filtering, and transporting fission products to the plant stack for elevated release. It accomplishes this purpose by maintaining a negative pressure of 0.25 inch of water within the reactor building as to the outside atmosphere to minimize unfiltered leakage of fission products from the reactor building and by exhausting filtered release of the primary and secondary containments through the ventilation stack. It also purges primary containment prior to outages when increased radioactivity is present and backs up the reactor building ventilation system for this function. During normal operation, the reactor building ventilation system is operating with the SGTS in standby. During a design basis accident (DBA), the SGTS fans are automatically started and effluents are filtered prior to release through the ventilation stack.

The SGTS contains safety-related components relied upon to remain functional during and following DBEs. In addition, the SGTS performs functions that support EQ.

The intended functions within the scope of license renewal include:

- provides mechanical closure
- provides pressure-retaining boundary; fission product barrier; containment isolation; or containment, holdup, and plateout (main steam system)
- provides flow restriction

In LRA Table 2.3.2.4, the applicant identified the following SGTS component types within the scope of license renewal and subject to an AMR:

- closure bolting
- damper housing
- door seal
- ductwork
- fan housing
- filter housing
- flexible connection
- flow element
- heater housing
- piping and fittings
- restricting orifice
- thermowell
- valve body

2.3.2.4.2 Staff Evaluation

The staff reviewed LRA Section 2.3.2.4 and UFSAR Sections 6.5.1, 7.3, 9.4.2, and 11.3.2.5 using the evaluation methodology of SER Section 2.3. The staff conducted its review in accordance with the guidance of SRP-LR Section 2.3.

In conducting its review, the staff evaluated the system functions described in the LRA and UFSAR to verify that the applicant had not omitted from the scope of license renewal any components with intended functions under 10 CFR 54.4(a). The staff then reviewed components that the applicant had identified as within the scope of license renewal to verify that it had not omitted any passive and long-lived components subject to an AMR in accordance with the requirements of 10 CFR 54.21(a)(1).

2.3.2.4.3 Conclusion

The staff reviewed the LRA to determine whether any SSCs that should be within the scope of license renewal had not been identified by the applicant. No omissions were identified. In addition, the staff determined whether any components subject to an AMR had not been identified by the applicant. No omissions were identified. The staff concludes that there is reasonable assurance that the applicant has adequately identified the SGTS components within the scope of license renewal, as required by 10 CFR 54.4(a), and those subject to an AMR, as required by 10 CFR 54.21(a)(1).

2.3.3 Auxiliary Systems

In LRA Section 2.3.3, the applicant identified the SCs of auxiliary systems subject to an AMR for license renewal.

The applicant described the supporting SCs of the auxiliary systems in the following sections of the LRA:

- 2.3.3.1   "C" battery room heating and ventilation
- 2.3.3.2   4160V switchgear room ventilation
- 2.3.3.3   480V switchgear room ventilation
- 2.3.3.4   battery and MG set room ventilation
- 2.3.3.5   chlorination system
- 2.3.3.6   circulating water system
- 2.3.3.7   containment inerting system
- 2.3.3.8   containment vacuum breakers
- 2.3.3.9   control rod drive system
- 2.3.3.10   control room HVAC
- 2.3.3.11   cranes and hoists
- 2.3.3.12   drywell floor and equipment drains
- 2.3.3.13   emergency diesel generator and auxiliary system
- 2.3.3.14   emergency service water system
- 2.3.3.15   fire protection system
- 2.3.3.16   fuel storage and handling equipment
- 2.3.3.17   hardened vent system
- 2.3.3.18   heating and process steam system
- 2.3.3.19   hydrogen and oxygen monitoring system
- 2.3.3.20   instrument (control) air system
- 2.3.3.21   main fuel oil storage and transfer system
- 2.3.3.22   miscellaneous floor and equipment drain system
- 2.3.3.23   nitrogen supply system
- 2.3.3.24   noble metals monitoring system
- 2.3.3.25   post-accident sampling system
- 2.3.3.26   process sampling system
- 2.3.3.27   radiation monitoring system
- 2.3.3.28   radwaste area heating and ventilation system
- 2.3.3.29   reactor building closed cooling water system
- 2.3.3.30   reactor building floor and equipment drains
- 2.3.3.31   reactor building ventilation system
- 2.3.3.32   reactor water cleanup system

- 2.3.3.33   roof drains and overboard discharge
- 2.3.3.34   sanitary waste system
- 2.3.3.35   service water system
- 2.3.3.36   shutdown cooling system
- 2.3.3.37   spent fuel pool cooling system
- 2.3.3.38   standby liquid control system (liquid poison system)
- 2.3.3.39   traveling in-core probe system
- 2.3.3.40   turbine building closed cooling water system
- 2.3.3.41   water treatment and distribution system

The staff's findings on LRA Sections 2.3.3.1 – 2.3.3.41 are presented in SER Sections 2.3.3.1 – 2.3.3.41, respectively.

2.3.3.1 "C" Battery Room Heating & Ventilation

2.3.3.1.1 Summary of Technical Information in the Application

In LRA Section 2.3.3.1, the applicant described the "C" battery room heating and ventilation system. The "C" battery room heating and ventilation system is a forced air ventilation system designed to maintain the "C" battery room within a specified temperature range and remove hydrogen produced by battery charging. This condition exists when the battery chargers are in operation and hydrogen is produced by the battery charging function. The "C" battery room ventilation system is a nonsafety-related system designed to support the 125V DC station "C" battery operation.

The failure of nonsafety-related SSCs in the "C" battery room heating and ventilation system potentially could prevent the satisfactory accomplishment of a safety-related function.

The intended functions within the scope of license renewal include:

- provides filtration
- provides mechanical closure
- provides pressure-retaining boundary; fission product barrier; containment isolation; or containment, holdup, and plateout (main steam system)

In LRA Table 2.3.3.1, the applicant identified the following "C" battery room heating and ventilation system component types within the scope of license renewal and subject to an AMR:

- bird screen
- closure bolting
- damper housing
- door seal
- ductwork
- fan housing
- filter housing
- flexible connection
- flow element
- louvers
- piping and fittings

2-53

2.3.3.1.2 Staff Evaluation

The staff reviewed LRA Section 2.3.3.1 and UFSAR Sections 8.3.2.1 and 9.4.3.2 using the evaluation methodology of SER Section 2.3. The staff conducted its review in accordance with the guidance of SRP-LR Section 2.3.

In conducting its review, the staff evaluated the system functions described in the LRA and UFSAR to verify that the applicant had not omitted from the scope of license renewal any components with intended functions under 10 CFR 54.4(a). The staff then reviewed components that the applicant had identified as within the scope of license renewal to verify that it had not omitted any passive and long-lived components subject to an AMR in accordance with the requirements of 10 CFR 54.21(a)(1).

2.3.3.1.3 Conclusion

The staff reviewed the LRA to determine whether any SSCs that should be within the scope of license renewal had not been identified by the applicant. No omissions were identified. In addition, the staff determined whether any components subject to an AMR had not been identified by the applicant. No omissions were identified. The staff concludes that there is reasonable assurance that the applicant has adequately identified the "C" battery room heating and ventilation system components within the scope of license renewal, as required by 10 CFR 54.4(a), and those subject to an AMR, as required by 10 CFR 54.21(a)(1).

2.3.3.2 4160V Switchgear Room Ventilation

2.3.3.2.1 Summary of Technical Information in the Application

In LRA Section 2.3.3.2, the applicant described the 4160V switchgear room ventilation system, a continuously operating forced air-flow system designed to remove heat produced by the operation of the switchgear and also to remove smoke in the event of a fire. The 4160V switchgear room ventilation system accomplishes this purpose by supplying the required air flow through the vaults necessary to keep the room temperatures within the design limits of the switchgear and to meet the smoke removal requirements of 10 CFR 50.48. The switchgear areas served by this ventilation system are in the turbine building within the 1C and 1D switchgear vaults. Each vault roof ventilation penetration is provided with a three-hour rated fire damper.

The 4160V switchgear room ventilation system performs functions that support fire protection and SBO.

The intended functions within the scope of license renewal include:

- provides filtration

- provides mechanical closure

- provides pressure-retaining boundary; fission product barrier; containment isolation; or containment, holdup, and plateout (main steam system)

In LRA Table 2.3.3.2, the applicant identified the following 4160V switchgear room ventilation system component types within the scope of license renewal and subject to an AMR:

- bird screen
- closure bolting
- damper housing
- fan housing

2.3.3.2.2 Staff Evaluation

The staff reviewed LRA Section 2.3.3.2 and UFSAR Section 9.4.3.2 using the evaluation methodology of SER Section 2.3. The staff conducted its review in accordance with the guidance of SRP-LR Section 2.3.

The staff reviewed the subsystems functions described in the LRA and UFSAR to verify that the applicant had not omitted from the scope of license renewal any components with intended functions under 10 CFR 54.4(a). The staff then reviewed components that the applicant had identified as within the scope of license renewal to verify that it had not omitted any passive and long-lived components subject to an AMR in accordance with the requirements of 10 CFR 54.21(a)(1).

2.3.3.2.3 Conclusion

The staff reviewed the LRA to determine whether any SSCs that should be within the scope of license renewal had not been identified by the applicant. No omissions were identified. In addition, the staff determined whether any components subject to an AMR had not been identified by the applicant. No omissions were identified. The staff concludes that there is reasonable assurance that the applicant has adequately identified the 4160V switchgear room ventilation system components within the scope of license renewal, as required by 10 CFR 54.4(a), and those subject to an AMR, as required by 10 CFR 54.21(a)(1).

2.3.3.3 480V Switchgear Room Ventilation

2.3.3.3.1 Summary of Technical Information in the Application

In LRA Section 2.3.3.3, the applicant described the 480V switchgear room ventilation system, a continuously operating forced air flow system designed to remove the heat produced by the operation of the 480V switchgear, and to also remove any smoke produced by a fire. The purpose of the system is to provide adequate ventilation to maintain the equipment environment within design temperature limits. The system accomplishes this purpose by utilizing supply and exhaust fans, a recirculation flow path, and ducting, dampers, and controls. The system consists of two independent ventilation trains, Train "A" for ventilation for 480V switchgear room A and train "B" for 480v switchgear room B. Train "A" also includes an alternate exhaust fan with intake and exhaust dampers. No heating or cooling is provided by this system.

The 480V switchgear room ventilation system contains safety-related components relied upon to remain functional during and following DBEs. In addition, the 480V switchgear room ventilation system performs functions that support fire protection and SBO.

The intended functions within the scope of license renewal include:

- provides filtration
- provides mechanical closure
- provides pressure-retaining boundary; fission product barrier; containment isolation; or containment, holdup, and plateout (main steam system)

In LRA Table 2.3.3.3, the applicant identified the following 480V switchgear room ventilation system component types within the scope of license renewal and subject to an AMR:

- bird screen
- closure bolting
- damper housing
- door seal
- ductwork
- fan housing
- filter housing
- flexible connection
- louvers
- piping and fittings
- sensor element
- valve body

2.3.3.3.2 Staff Evaluation

The staff reviewed LRA Section 2.3.3.3 and UFSAR Section 9.4.5.2.6 using the evaluation methodology of SER Section 2.3. The staff conducted its review in accordance with the guidance of SRP-LR Section 2.3.

The staff reviewed the subsystems functions described in the LRA and UFSAR to verify that the applicant had not omitted from the scope of license renewal any components with intended functions under 10 CFR 54.4(a). The staff then reviewed components that the applicant had identified as within the scope of license renewal to verify that it had not omitted any passive and long-lived components subject to an AMR in accordance with the requirements of 10 CFR 54.21(a)(1).

2.3.3.3.3 Conclusion

The staff reviewed the LRA to determine whether any SSCs that should be within the scope of license renewal had not been identified by the applicant. No omissions were identified. In addition, the staff determined whether any components subject to an AMR had not been identified by the applicant. No omissions were identified. The staff concludes that there is reasonable assurance that the applicant has adequately identified the 480V switchgear room ventilation system components within the scope of license renewal, as required by 10 CFR 54.4(a), and those subject to an AMR, as required by 10 CFR 54.21(a)(1).

2.3.3.4 Battery and MG Set Room Ventilation

2.3.3.4.1 Summary of Technical Information in the Application

In LRA Section 2.3.3.4, the applicant described the battery and motor generator (MG) set room ventilation system, a continuously operating forced air flow system designed to remove the heat produced by operating equipment. The system is also designed to remove gasses produced by the A and B station batteries and to remove any smoke produced by a fire. The purpose of the system is to provide adequate ventilation to maintain the equipment environment within design temperature limits and to remove any hydrogen released from the batteries. The system is supplemented with an air conditioning unit to provide additional MG set cooling when required. The system accomplishes this purpose by utilizing supply and exhaust fans, a recirculation flow path, and an air conditioning unit with ducting, dampers, and controls. The supply system flow splits to supply both the battery room and the MG room, and the exhaust system draws air from both rooms. This system is actuated when the motor approaches or exceeds a set temperature. The system is manually initiated and normally in operation.

The battery and MG set room ventilation system contains safety-related components relied upon to remain functional during and following DBEs. In addition, the battery and MG set room ventilation system performs functions that support fire protection and SBO.

The intended functions within the scope of license renewal include:

*   provides filtration

*   provides mechanical closure

*   provides pressure-retaining boundary; fission product barrier; containment isolation; or containment, holdup, and plateout (main steam system)

In LRA Table 2.3.3.4, the applicant identified the following battery and MG set room ventilation system component types within the scope of license renewal and subject to an AMR:

*   bird screen
*   closure bolting
*   damper housing
*   door seal
*   ductwork
*   fan housing
*   filter housing
*   flexible connection
*   flow element (pitot tube)
*   louvers
*   piping and fittings
*   sensor element (temperature)
*   valve body

2.3.3.4.2 Staff Evaluation

The staff reviewed LRA Section 2.3.3.4 and UFSAR Section 9.4.5.2.5 using the evaluation methodology of SER Section 2.3. The staff conducted its review in accordance with the guidance of SRP-LR Section 2.3.

The staff reviewed the subsystems functions described in the LRA and UFSAR to verify that the applicant had not omitted from the scope of license renewal any components with intended functions under 10 CFR 54.4(a). The staff then reviewed components that the applicant had identified as within the scope of license renewal to verify that it had not omitted any passive and long-lived components subject to an AMR in accordance with the requirements of 10 CFR 54.21(a)(1).

2.3.3.4.3 Conclusion

The staff reviewed the LRA to determine whether any SSCs that should be within the scope of license renewal had not been identified by the applicant. No omissions were identified. In addition, the staff determined whether any components subject to an AMR had not been identified by the applicant. No omissions were identified. The staff concludes that there is reasonable assurance that the applicant has adequately identified the battery and MG set room ventilation system components within the scope of license renewal, as required by 10 CFR 54.4(a), and those subject to an AMR, as required by 10 CFR 54.21(a)(1).

2.3.3.5 Chlorination System

2.3.3.5.1 Summary of Technical Information in the Application

In LRA Section 2.3.3.5, the applicant described the chlorination system, which operates year-round and is designed to inject sodium hypochlorite to various points in the circulating water, service water, and emergency service water systems. The purpose of the system is to eliminate or reduce biofouling while maintaining residual chlorine concentration at the discharge canal within federal and state regulations. The system accomplishes the purpose by treatment of systems using bay water as a heat sink in order to minimize micro and macro biofouling of heat exchangers. Biofouling, if left unchecked, will affect performance. It accomplishes this check by chlorine bonding with amines in the marine environment to form toxic chloramine compounds. It also displaces bromine and iodine, both essential marine salts. Marine life, dependent upon a stable balance of chemistry, dies. The chlorination system is comprised of two hypochlorite storage tanks, two eductors, and the required piping, valves, instrumentation, and controls. The sodium hypochlorite is stored in two 6500-gallon plastic storage tanks. The system is located within the chlorination building and adjacent pad with the exception of the piping routed below grade and in the turbine building.

The failure of nonsafety-related SSCs in the chlorination system potentially could prevent the satisfactory accomplishment of a safety-related function.

The intended functions within the scope of license renewal include:

- maintains mechanical and structural integrity to prevent spatial interactions that could cause failure of safety-related SSCs (includes the required structural support when the nonsafety-related leakage boundary piping is also attached to safety-related piping)
- provides mechanical closure

In LRA Table 2.3.3.5, the applicant identified the following chlorination system component types within the scope of license renewal and subject to an AMR:

- closure bolting
- piping and fittings
- valve body

2.3.3.5.2 Staff Evaluation

The staff reviewed LRA Section 2.3.3.5 and UFSAR Section 10.4.5.2 using the evaluation methodology of SER Section 2.3. The staff conducted its review in accordance with the guidance of SRP-LR Section 2.3.

In conducting its Tier-1 review of the BOP two-tier review process, the staff evaluated the system functions described in the LRA and UFSAR to verify that the applicant had not omitted from the scope of license renewal any components with intended functions under 10 CFR 54.4(a). The staff then reviewed those components that the applicant had identified as within the scope of license renewal to verify that it had not omitted any passive and long-lived components subject to an AMR in accordance with the requirements of 10 CFR 54.21(a)(1).

2.3.3.5.3 Conclusion

The staff reviewed the LRA to determine whether any SSCs that should be within the scope of license renewal had not been identified by the applicant. No omissions were identified. In addition, the staff determined whether any components subject to an AMR had not been identified by the applicant. No omissions were identified. The staff concludes that there is reasonable assurance that the applicant has adequately identified the chlorination system components within the scope of license renewal, as required by 10 CFR 54.4(a), and those subject to an AMR, as required by 10 CFR 54.21(a)(1).

2.3.3.6 Circulating Water System

2.3.3.6.1 Summary of Technical Information in the Application

In LRA Section 2.3.3.6, the applicant described the circulating water system (CWS), a low-pressure, high-volume open-cycle cooling water system designed to provide cooling water to the main condenser and the main source of cooling water for the turbine building closed cooling water (TBCCW) heat exchangers. If TBCCW heat exchanger cooling water is not available from the CWS, the service water system (SWS) provides the cooling water to the TBCCW heat exchangers. The CWS pumps are located at the intake structure in separate chambers. The pumps draw sea water from the intake canal and discharge the water into large diameter pipe lines that deliver the cooling water to the intake tunnel. Each pump discharge line has an isolation valve and local pressure instrumentation. From the intake tunnel the water flows into large individual pipes that supply the cooling water to each condenser shell. Each of these cooling water supply lines has an isolation valve and a chlorination system connection. Heat is absorbed by the cooling water, increasing the water discharge temperature. The heated water is discharged through large lines to the discharge tunnel. Each discharge line has an isolation valve. The discharge tunnel delivers the water to the discharge canal and the water flows from the canal into Barnegat Bay. Deicing recirculation is provided during cold weather operation.

The failure of nonsafety-related SSCs in the CWS potentially could prevent the satisfactory accomplishment of a safety-related function.

The intended functions within the scope of license renewal include:

- maintains mechanical and structural integrity to prevent spatial interactions that could cause failure of safety-related SSCs (includes the required structural support when the nonsafety-related leakage boundary piping is also attached to safety-related piping)
- provides mechanical closure

In LRA Table 2.3.3.6, the applicant identified the following CWS component types within the scope of license renewal and subject to an AMR:

- closure bolting
- expansion joint
- flow glass
- flow indicator
- level glass
- piping and fittings
- strainer body
- thermowell
- valve body

2.3.3.6.2 Staff Evaluation

The staff reviewed LRA Section 2.3.3.6 and UFSAR Section 10.4.5 using the evaluation methodology of SER Section 2.3. The staff conducted its review in accordance with the guidance of SRP-LR Section 2.3.

In conducting its Tier-2 review of the BOP two-tier review process, the staff evaluated the system functions described in the LRA and UFSAR to verify that the applicant had not omitted from the scope of license renewal any components with intended functions under 10 CFR 54.4(a). The staff then reviewed those components that the applicant had identified as within the scope of license renewal to verify that it had not omitted any passive and long-lived components subject to an AMR in accordance with the requirements of 10 CFR 54.21(a)(1).

2.3.3.6.3 Conclusion

The staff reviewed the LRA to determine whether any SSCs that should be within the scope of license renewal had not been identified by the applicant. No omissions were identified. In addition, the staff determined whether any components subject to an AMR had not been identified by the applicant. No omissions were identified. The staff concludes that there is reasonable assurance that the applicant has adequately identified the CWS components within the scope of license renewal, as required by 10 CFR 54.4(a), and those subject to an AMR, as required by 10 CFR 54.21(a)(1).

2.3.3.7 Containment Inerting System

2.3.3.7.1 Summary of Technical Information in the Application

In LRA Section 2.3.3.7, the applicant described the containment inerting system (CIS), a pressurized gas system designed to maintain an inert atmosphere within the primary containment to preclude energy releases from a possible hydrogen-oxygen reaction following a postulated LOCA. The purpose of the CIS is to provide primary containment purging and makeup in order to control the oxygen concentration inside the primary containment. To ready the primary containment for power operation, the CIS accomplishes the purpose of purging by introducing nitrogen to displace the oxygen from the free volume in the primary containment. During power operation, the CIS accomplishes the purpose of makeup by introducing nitrogen to maintain a low oxygen concentration in the primary containment. During power operation, when nitrogen makeup is not in service, the nitrogen atmosphere is isolated within the primary containment and recirculated by the drywell cooling system. Following a DBA LOCA, the CIS accomplishes the purpose of purging by introducing nitrogen into the primary containment to control post-LOCA hydrogen and oxygen concentrations to below combustible levels. CIS operation in both the purge and makeup modes is initiated manually. The CIS receives vaporized nitrogen through two headers from the nitrogen supply system, the purge header and the nitrogen makeup header.

The CIS contains safety-related components relied upon to remain functional during and following DBEs. In addition, the CIS performs functions that support fire protection and EQ.

The intended functions within the scope of license renewal include:

- provides mechanical closure
- provides pressure-retaining boundary; fission product barrier; containment isolation; or containment, holdup, and plateout (main steam system)

In LRA Table 2.3.3.7, the applicant identified the following CIS component types within the scope of license renewal and subject to an AMR:

- closure bolting
- drain trap
- flow element
- piping and fittings
- thermowell
- valve body

2.3.3.7.2 Staff Evaluation

The staff reviewed LRA Section 2.3.3.7 and UFSAR Section 6.2.5 using the evaluation
methodology of SER Section 2.3. The staff conducted its review in accordance with the
guidance of SRP-LR Section 2.3.

The staff reviewed the system functions described in the LRA and UFSAR to verify that the
applicant had not omitted from the scope of license renewal any components with intended
functions under 10 CFR 54.4(a). The staff then reviewed components that the applicant had
identified as within the scope of license renewal to verify that it had not omitted any passive and
long-lived components subject to an AMR in accordance with the requirements of
10 CFR 54.21(a)(1).

2.3.3.7.3 Conclusion

The staff reviewed the LRA to determine whether any SSCs that should be within the scope of
license renewal had not been identified by the applicant. No omissions were identified. In
addition, the staff determined whether any components subject to an AMR had not been
identified by the applicant. No omissions were identified. The staff concludes that there is
reasonable assurance that the applicant has adequately identified the CIS components within
the scope of license renewal, as required by 10 CFR 54.4(a), and those subject to an AMR, as
required by 10 CFR 54.21(a)(1).

2.3.3.8 Containment Vacuum Breakers

2.3.3.8.1 Summary of Technical Information in the Application

In LRA Section 2.3.3.8, the applicant described the containment vacuum breaker (CVB) system,
two systems designed to prevent torus water from backing up into the drywell during various
reactor leakage and suppression condensation modes and limit negative pressure differentials
on the drywell in conjunction with the reactor building to torus vacuum relief system. These
systems are the torus to drywell and the reactor building to torus vacuum relief systems. The
purpose of the torus to drywell vacuum relief system is to prevent the drywell pressure from
dropping significantly below the pressure in the torus airspace. The reactor building to torus
vacuum relief system is intended to prevent the torus air space pressure from dropping
significantly below the ambient atmospheric pressure in the reactor building. The reactor building
to torus vacuum breakers accomplish their purpose by opening automatically at a predetermined
differential pressure. The torus to drywell vacuum breakers accomplish theirs by venting
non-condensable gas (carryover to the torus during an accident) back to the drywell from the
torus. The primary containment has a vacuum breaker system to equalize the pressure between
the drywell and the torus and between the torus and the reactor building. The CVB system
assures that the external design pressure limits of the two chambers are not exceeded.

The CVB system contains safety-related components relied upon to remain functional during
and following DBEs. In addition, the CVB system performs functions that support fire protection
and EQ.

The intended functions within the scope of license renewal include:

* provides mechanical closure

- provides pressure-retaining boundary; fission product barrier; containment isolation; or containment, holdup, and plateout (main steam system)

In LRA Table 2.3.3.8, the applicant identified the following CVB system component types within the scope of license renewal and subject to an AMR:

- closure bolting
- expansion joint
- piping and fittings
- valve body
- valve body (vacuum breakers)

2.3.3.8.2 Staff Evaluation

The staff reviewed LRA Section 2.3.3.8 and UFSAR Section 6.2.2 using the evaluation methodology of SER Section 2.3. The staff conducted its review in accordance with the guidance of SRP-LR Section 2.3.

The staff reviewed the system functions described in the LRA and UFSAR to verify that the applicant had not omitted from the scope of license renewal any components with intended functions under 10 CFR 54.4(a). The staff then reviewed components that the applicant had identified as within the scope of license renewal to verify that it had not omitted any passive and long-lived components subject to an AMR in accordance with the requirements of 10 CFR 54.21(a)(1).

2.3.3.8.3 Conclusion

The staff reviewed the LRA to determine whether any SSCs that should be within the scope of license renewal had not been identified by the applicant. No omissions were identified. In addition, the staff determined whether any components subject to an AMR had not been identified by the applicant. No omissions were identified. The staff concludes that there is reasonable assurance that the applicant has adequately identified the CVB system components within the scope of license renewal, as required by 10 CFR 54.4(a), and those subject to an AMR, as required by 10 CFR 54.21(a)(1).

2.3.3.9 Control Rod Drive System

2.3.3.9.1 Summary of Technical Information in the Application

In LRA Section 2.3.3.9, the applicant described the CRD system, the primary purpose of which is to rapidly insert negative reactivity to shut down the reactor under accident or transient conditions and to manage reactivity in the reactor core by inserting or withdrawing control rods at a limited rate, one rod at a time, for power level control and flux shaping during normal reactor operation. The CRD system accomplishes this purpose by providing water at the required operating pressures to the control rod drives for cooling and for all types of control rod motion in response to inputs from the reactor manual control system (RMCS) and RPS. The secondary purpose of the CRD system is to supply the reactor head cooling system (RHCS). It accomplishes this purpose by providing water at the required pressure to the reactor vessel head spray nozzle used to cool the upper head region during plant cooldown. The CRD system is comprised of CRD mechanisms and the CRD hydraulic system. Each of the CRDMs is a double-acting, mechanically-latched, hydraulic cylinder with reactor grade water as the operating fluid. Each CRD mechanism is capable of inserting or withdrawing the attached control rod at a slow, controlled rate as well as rapidly in an emergency. A locking mechanism allows a drive to be positioned during stroking to hold the control rod in a fixed position.

The CRD system contains safety-related components relied upon to remain functional during and following DBEs. The failure of nonsafety-related SSCs in the CRD system potentially could prevent the satisfactory accomplishment of a safety-related function. In addition, the CRD system performs functions that support fire protection, SBO, and EQ.

The intended functions within the scope of license renewal include:

- provides filtration
- maintains mechanical and structural integrity to prevent spatial interactions that could cause failure of safety-related SSCs (includes the required structural support when the nonsafety-related leakage boundary piping is also attached to safety-related piping)
- provides mechanical closure
- provides pressure-retaining boundary; fission product barrier; containment isolation; or containment, holdup, and plateout (main steam system)
- provides flow restriction

In LRA Table 2.3.3.9, the applicant identified the following CRD system component types within the scope of license renewal and subject to an AMR:

- accumulator
- closure bolting
- filter
- filter housing
- flow element
- gauge snubber
- gear box
- piping and fittings
- pump casing
- restricting orifice
- rupture disks
- strainer
- strainer body
- valve body

2.3.3.9.2 Staff Evaluation

The staff reviewed LRA Section 2.3.3.9 and UFSAR Sections 3.9.4, 4.5, 4.6, and 15.8 using the evaluation methodology of SER Section 2.3. The staff conducted its review in accordance with the guidance of SRP-LR Section 2.3.

In conducting its review, the staff evaluated the system functions described in the LRA and UFSAR to verify that the applicant had not omitted from the scope of license renewal any components with intended functions under 10 CFR 54.4(a). The staff then reviewed those components that the applicant had identified as within the scope of license renewal to verify that it had not omitted any passive and long-lived components subject to an AMR in accordance with the requirements of 10 CFR 54.21(a)(1).

2.3.3.9.3 Conclusion

The staff reviewed the LRA to determine whether any SSCs that should be within the scope of license renewal had not been identified by the applicant. No omissions were identified. In addition, the staff determined whether any components subject to an AMR had not been identified by the applicant. No omissions were identified. The staff concludes that there is reasonable assurance that the applicant has adequately identified the CRD system components within the scope of license renewal, as required by 10 CFR 54.4(a), and those subject to an AMR, as required by 10 CFR 54.21(a)(1).

2.3.3.10 Control Room HVAC

2.3.3.10.1 Summary of Technical Information in the Application

In LRA Section 2.3.3.10, the applicant described the control room heating, ventilation, and air conditioning (HVAC) system that serves the control room envelope, which consists of the control room and lower cable spreading room. The control room HVAC system is evaluated with the separate miscellaneous HVAC license renewal system. The purpose of the control room HVAC system is to maintain a comfortable temperature and provide ventilation for personnel and equipment during normal operation. It also incorporates three incident modes of operation to provide a habitable environment for control room operators and equipment cooling after radiological releases from DBAs during or after toxic chemical releases and for fires inside the control room. The normally operating system is initiated into incident modes manually. In addition to normal operation, three incident modes of partial recirculation, full recirculation, and purge are available. In the event of a DBA manual selection of the partial recirculation mode maintains the control room envelope at a positive pressure with minimal infiltration. During toxic gas releases, the full recirculation mode uses no outside air for minimal intrusion of toxic gases. In the event of smoke in the control room envelope, purge mode selection supplies all outdoor air to avoid recirculation and clear smoke and fumes.

The control room HVAC system contains safety-related components relied upon to remain functional during and following DBEs. In addition, the control room HVAC system performs functions that support fire protection and SBO.

The intended functions within the scope of license renewal include:

- provides filtration
- provides heat transfer
- provides mechanical closure
- provides pressure-retaining boundary; fission product barrier; containment isolation; or containment, holdup, and plateout (main steam system)

2-65

In LRA Table 2.3.3.10, the applicant identified the following control room HVAC system component types within the scope of license renewal and subject to an AMR:

- bird screen
- closure bolting
- damper housing
- door seal
- ductwork
- fan housing
- filter housing
- flexible connection
- heat exchangers (condensing coil)
- heat exchangers (evaporator coil)
- heater housing
- louvers
- piping and fittings

2.3.3.10.2 Staff Evaluation

The staff reviewed LRA Section 2.3.3.10 and UFSAR Sections 9.4.1, 6.4.1, and 12.3.3 using the evaluation methodology of SER Section 2.3. The staff conducted its review in accordance with the guidance of SRP-LR Section 2.3.

The staff reviewed the subsystems functions described in the LRA and UFSAR to verify that the applicant had not omitted from the scope of license renewal any components with intended functions under 10 CFR 54.4(a). The staff then reviewed components that the applicant had identified as within the scope of license renewal to verify that it had not omitted any passive and long-lived components subject to an AMR in accordance with the requirements of 10 CFR 54.21(a)(1).

2.3.3.10.3 Conclusion

The staff reviewed the LRA to determine whether any SSCs that should be within the scope of license renewal had not been identified by the applicant. No omissions were identified. In addition, the staff determined whether any components subject to an AMR had not been identified by the applicant. No omissions were identified. The staff concludes that there is reasonable assurance that the applicant has adequately identified the control room HVAC system components within the scope of license renewal, as required by 10 CFR 54.4(a), and those subject to an AMR, as required by 10 CFR 54.21(a)(1).

2.3.3.11 Cranes and Hoists

2.3.3.11.1 Summary of Technical Information in the Application

In LRA Section 2.3.3.11, the applicant described the cranes and hoists system comprised of load handling overhead bridge cranes, monorails, jib cranes, and hoists throughout the facility to support operation and maintenance activities. The system includes cranes and hoists required to comply with the requirements of NUREG-0612, "Control of Heavy Loads," and hoists for handling light load. Major cranes include the reactor building and the turbine building cranes. The reactor building crane services the operating floor and is used to lift all heavy loads that must travel over the spent fuel pool. The crane is also used to handle new fuel and transport the spent fuel cask and has been upgraded to a single failure-proof criterion in accordance with NUREG-0612 and NUREG-0554. The turbine building crane handles heavy loads in the turbine building, primarily supporting turbine repairs or maintenance. Included in the evaluation boundary of cranes and hoists system are load handling systems in various areas of the facility. Cranes and hoists are classified non-safety related and designed to seismic Class II criteria.

The cranes and hoists system contains safety-related components relied upon to remain functional during and following DBEs. The failure of nonsafety-related SSCs in the cranes and hoists system potentially could prevent the satisfactory accomplishment of a safety-related function.

The intended function, within the scope of license renewal, is to provide structural support or structural integrity to preclude nonsafety-related component interactions that could prevent satisfactory accomplishment of a safety-related function.

In LRA Table 2.3.3-11, the applicant identified the following cranes and hoists system component types within the scope of license renewal and subject to an AMR:

- crane (bridge; trolley)
- crane (bridge; trolley; girders)
- jib cranes (columns; beams; anchorage)
- monorails, and hoists (beams; plates)
- rail system (rail, plates, clips)
- structural bolts

2.3.3.11.2 Staff Evaluation

The staff reviewed LRA Section 2.3.3.11 and UFSAR Section 9.1.4.2.3 using the evaluation methodology of SER Section 2.3. The staff conducted its review in accordance with the guidance of SRP-LR Section 2.3.

In conducting its Tier-1 review of the BOP two-tier review process, the staff evaluated the system functions described in the LRA and UFSAR to verify that the applicant had not omitted from the scope of license renewal any components with intended functions under 10 CFR 54.4(a). The staff then reviewed those components that the applicant had identified as within the scope of license renewal to verify that it had not omitted any passive and long-lived components subject to an AMR in accordance with the requirements of 10 CFR 54.21(a)(1).

2.3.3.11.3 Conclusion

The staff reviewed the LRA to determine whether any SSCs that should be within the scope of license renewal had not been identified by the applicant. No omissions were identified. In addition, the staff determined whether any components subject to an AMR had not been identified by the applicant. No omissions were identified. The staff concludes that there is reasonable assurance that the applicant has adequately identified the cranes and hoists system components within the scope of license renewal, as required by 10 CFR 54.4(a), and those subject to an AMR, as required by 10 CFR 54.21(a)(1).

2.3.3.12 Drywell Floor and Equipment Drains

2.3.3.12.1 Summary of Technical Information in the Application

In LRA Section 2.3.3.12, the applicant described the drywell floor and equipment drains (DFEDs) comprised of both gravity and pumped fluid lines designed for the collection of drainage from floor and equipment drains located in the drywell structure and transfer of the drainage to the radwaste system. They also include that portion of the RCPB leak detection function comprised of the instrumentation monitoring the drywell floor drain sump fill time and pump flow rates from the drywell floor drain sump and drywell equipment drain tank. The DFED accomplish this purpose by collecting floor drainage and condensed steam from the drywell air coolers in the drywell floor drain sump and equipment drainage in the drywell equipment drain tank and using submersible pumps from the sump and duplex pumps from the drain tank to transfer the collected drainage to radwaste system collection tanks for processing. Both identified and unidentified leakage are collected by the DFEDs.

The DFEDs contain safety-related components relied upon to remain functional during and following DBEs. The failure of nonsafety-related SSCs in the DFEDs potentially could prevent the satisfactory accomplishment of a safety-related function. In addition, the DFEDs performs functions that support EQ.

The intended functions within the scope of license renewal include:

- maintains mechanical and structural integrity to prevent spatial interactions that could cause failure of safety-related SSCs (includes the required structural support when the nonsafety-related leakage boundary piping is also attached to safety-related piping)
- provides mechanical closure
- provides pressure-retaining boundary
- provides structural support or structural integrity to preclude nonsafety-related component interactions that could prevent satisfactory accomplishment of a safety-related function

In LRA Table 2.3.3.12, the applicant identified the following DFEDs component types within the scope of license renewal and subject to an AMR:

- closure bolting
- flow element
- flow glass
- heat exchanger
- piping and fittings
- pump casing
- tanks
- valve body

2.3.3.12.2 Staff Evaluation

The staff reviewed LRA Section 2.3.3.12 and UFSAR Sections 5.2.5, 9.3.3, and 11.2 using the evaluation methodology of SER Section 2.3. The staff conducted its review in accordance with the guidance of SRP-LR Section 2.3.

In conducting its Tier-2 review of the BOP two-tier review process, the staff evaluated the system functions described in the LRA and UFSAR to verify that the applicant had not omitted from the scope of license renewal any components with intended functions under 10 CFR 54.4(a). The staff then reviewed those components that the applicant had identified as within the scope of license renewal to verify that it had not omitted any passive and long-lived components subject to an AMR in accordance with the requirements of 10 CFR 54.21(a)(1).

2.3.3.12.3 Conclusion

The staff reviewed the LRA to determine whether any SSCs that should be within the scope of license renewal had not been identified by the applicant. No omissions were identified. In addition, the staff determined whether any components subject to an AMR had not been identified by the applicant. No omissions were identified. The staff concludes that there is reasonable assurance that the applicant has adequately identified the DFEDs components within the scope of license renewal, as required by 10 CFR 54.4(a), and those subject to an AMR, as required by 10 CFR 54.21(a)(1).

2.3.3.13 *Emergency Diesel Generator and Auxiliary System*

2.3.3.13.1 Summary of Technical Information in the Application

In LRA Section 2.3.3.13, the applicant described the EDG and auxiliary system, the purpose of which is to provide sufficient power independently to energize all equipment required for safely shutting down the reactor. It accomplishes this purpose using two diesel generator units located in separate rooms of a stand-alone, reinforced concrete structure. Each diesel engine powers a generator at a voltage compatible to the plant electrical distribution systems with sufficient output capacity to meet plant shutdown loads. Each diesel generator is equipped with its own starting system, cooling system, lubrication system, combustion air and equipment cooling system, a fuel oil storage and transfer system, and all the auxiliaries that allow it to perform its function. The diesels are automatically started by a reactor low-low level, a high drywell pressure signal, by an undervoltage condition in the 4160V AC system, or by a low diesel generator lube oil temperature. The diesels can be remotely manually started from the control room or at the local EDG switchgear panels.

The EDG and auxiliary system contains safety-related components relied upon to remain functional during and following DBEs. The failure of nonsafety-related SSCs in the EDG and auxiliary system potentially could prevent the satisfactory accomplishment of a safety-related function. In addition, the EDG and auxiliary system performs functions that support fire protection and SBO.

The intended functions within the scope of license renewal include:

*   provides spray shield or curbs for directing flow

- provides filtration
- provides rated fire barrier
- provides heat transfer
- maintains mechanical and structural integrity to prevent spatial interactions that could cause failure of safety-related SSCs (includes the required structural support when the nonsafety-related leakage boundary piping is also attached to safety-related piping)
- provides mechanical closure
- provides pressure-retaining boundary
- provides flow restriction

In LRA Table 2.3.3.13, the applicant identified the following EDG and auxiliary system component types within the scope of license renewal and subject to an AMR:

- bird screen
- closure bolting
- ductwork
- exhaust stack
- fan housing (dust bin blower fan)
- fan housing (radiator fan)
- filter (inertial air bin)
- filter (oil bath)
- filter housing (air cooling)
- filter housing (fuel oil)
- filter housing (lube oil)
- flame arrester (fuel oil tank)
- flexible hose
- heat exchanger (lube oil cooler)
- heat exchangers (radiator)
- louvers
- muffler
- piping and fittings
- pump casing (fuel oil)
- pump casing (lube oil)
- restricting orifice
- sensor element (lube oil)
- sensor element (temperature control manifold)
- sight glasses
- strainer
- strainer body
- tanks (fuel day tank)
- tanks (fuel oil tank)
- tanks (immersion heater)
- tanks (water tank)
- temperature control manifold (water cooling)
- thermowell
- valve body

2.3.3.13.2 Staff Evaluation

The staff reviewed LRA Section 2.3.3.13 and UFSAR Sections 8.3.1.1.5, 9.5.4, 9.5.5, 9.5.6, 9.5.7, 9.5.8, and 9.5.9 using the evaluation methodology of SER Section 2.3. The staff conducted its review in accordance with the guidance of SRP-LR Section 2.3.

In conducting its Tier-2 review of the BOP two-tier review process, the staff evaluated the system functions described in the LRA and UFSAR to verify that the applicant had not omitted from the scope of license renewal any components with intended functions under 10 CFR 54.4(a). The staff then reviewed those components that the applicant had identified as within the scope of license renewal to verify that it had not omitted any passive and long-lived components subject to an AMR in accordance with the requirements of 10 CFR 54.21(a)(1).

2.3.3.13.3 Conclusion

The staff reviewed the LRA to determine whether any SSCs that should be within the scope of license renewal had not been identified by the applicant. No omissions were identified. In addition, the staff determined whether any components subject to an AMR had not been identified by the applicant. No omissions were identified. The staff concludes that there is reasonable assurance that the applicant has adequately identified the EDG and auxiliary system components within the scope of license renewal, as required by 10 CFR 54.4(a), and those subject to an AMR, as required by 10 CFR 54.21(a)(1).

2.3.3.14 Emergency Service Water System

2.3.3.14.1 Summary of Technical Information in the Application

In LRA Section 2.3.3.14, the applicant described the ESW system, which, along with the containment spray system, comprise the containment heat removal systems. The purpose of this system is to aid the containment spray system in removing fission product decay heat from the primary containment following a design-basis LOCA. This system is also used during normal operation to cool the torus when necessary. It accomplishes this purpose by supplying cooling water, from the ultimate heat sink (intake canal), to the containment spray heat exchangers and transferring the heat energy to the environment via the discharge canal. During normal plant operations, when ESW is in standby, the SWS supplies a constant flow of water through the containment spray heat exchangers to maintain them full of chlorinated water. Sodium hypochlorite is injected into the ESW system via the SWS keep fill line. Additionally, ESW can be cross-connected with the SWS to allow ESW to provide an alternate cooling path during plant shutdown and during SWS maintenance.

The ESW system contains safety-related components relied upon to remain functional during and following DBEs. The failure of nonsafety-related SSCs in the ESW system potentially could prevent the satisfactory accomplishment of a safety-related function. In addition, the ESW system performs functions that support fire protection and EQ.

The intended functions within the scope of license renewal include:

- provides heat transfer
- maintains mechanical and structural integrity to prevent spatial interactions that could cause failure of safety-related SSCs (includes the required structural support when the nonsafety-related leakage boundary piping is also attached to safety-related piping)
- provides mechanical closure
- provides pressure-retaining boundary
- provides flow restriction

In LRA Table 2.3.3.14, the applicant identified the following ESW system component types within the scope of license renewal and subject to an AMR:

- closure bolting
- expansion joint
- flow element
- heat exchangers (containment spray)
- piping and fittings
- pump casing (ESW pumps)
- pump casing (HTXR drain pumps)
- restricting orifice
- sight glasses
- thermowell
- valve body

2.3.3.14.2 Staff Evaluation

The staff reviewed LRA Section 2.3.3.14 and UFSAR Sections 6.2.2, 7.3.1 and 9.2.1 using the evaluation methodology of SER Section 2.3. The staff conducted its review in accordance with the guidance of SRP-LR Section 2.3.

In conducting its Tier-2 review of the BOP two-tier review process, the staff evaluated the system functions described in the LRA and UFSAR to verify that the applicant had not omitted from the scope of license renewal any components with intended functions under 10 CFR 54.4(a). The staff then reviewed those components that the applicant had identified as within the scope of license renewal to verify that it had not omitted any passive and long-lived components subject to an AMR in accordance with the requirements of 10 CFR 54.21(a)(1).

The staff of LRA Section 2.3.3.14 identified an area in which additional information was necessary to complete the review of the applicant's scoping and screening results. The applicant responded to the staff's RAI as discussed below.

In RAI 2.3.3.14-1 dated December 28, 2005, the staff stated that several strainers not identified on LRA Table 2.3.3.14 as requiring aging management are indicated as within the scope of license renewal on its license renewal drawing. The staff requested that the applicant clarify whether these long-lived passive components are subject to an AMR or justify their exclusion from LRA Table 2.3.3.14.

In its response dated January 26, 2006, the applicant stated:

> The strainer symbols shown on license renewal drawing LR-BR-2005, Sheet 4 at drawing coordinates C-7, C-8, F-7, and F-8 are depicting the diaphragm seal that is integral to the pressure indicator assembly. The diaphragm seal is not specifically called out in LRA Table 2.3.3.14 since it is considered part of the "active" pressure instrument. Diaphragm seals isolate pressure instruments from the process media while allowing the instrument to sense the process pressure. A diaphragm, together with a fill fluid, transmits pressure from the process medium to the pressure element assembly of the instrument. There would be no need to filter the medium prior to the diaphragm seal.
>
> Because these diaphragm seals are part of the pressure indicator assembly, which is an "active" component, they are not subject to aging management review.

The staff finds the applicant's response acceptable because it adequately clarified that the components in question are active (parts of an instrument assembly) and not subject to an AMR under 10 CFR 54.21(a)(1). The staff's concern described in RAI 2.3.3.14-1 is resolved.

2.3.3.14.3 Conclusion

The staff reviewed the LRA and the RAI response to determine whether any SSCs that should be within the scope of license renewal had not been identified by the applicant. No omissions were identified. In addition, the staff determined whether any components subject to an AMR had not been identified by the applicant. No omissions were identified. The staff concludes that there is reasonable assurance that the applicant has adequately identified the ESW system components within the scope of license renewal, as required by 10 CFR 54.4(a), and those subject to an AMR, as required by 10 CFR 54.21(a)(1).

2.3.3.15 Fire Protection System

2.3.3.15.1 Summary of Technical Information in the Application

In LRA Section 2.3.3.15, the applicant described the fire protection system, a normally operating mechanical system designed to provide for the rapid detection and suppression of a fire at the plant. The purpose of the fire protection system is to promptly detect, contain, and extinguish fires if they occur, maintain the capability to safely shut down the plant if fires occur, and prevent the release of a significant amount of radiation in the event of a fire. The fire protection system accomplishes this purpose by providing fire protection in the form of detection, alarms, fire barriers, and suppression for selected areas of the plant. The fire protection system consists of the fire protection water system, carbon dioxide (CO_2) gas systems, halon systems, portable foam equipment, portable fire extinguishers, and fire detection and signaling systems. These systems work in conjunction with physical plant design features to provide overall fire protection for OCGS. The physical plant design features consist of fire barrier walls and slabs, fire barrier penetration seals, fire doors, fire-rated enclosures (including steel fire wrap), and dikes credited for containing oil spills.

The fire protection system performs an intended function for compliance with fire protection regulations. The fire protection system works in conjunction with fire barriers and other plant design features and established safe-shutdown systems and procedures for compliance with 10 CFR 50.48. The failure of nonsafety-related SSCs in the fire protection system potentially could prevent the satisfactory accomplishment of a safety-related function.

The intended functions within the scope of license renewal include:

- provides filtration

- provides rated fire barrier (dikes to contain oil spill)

- provides rated fire barrier (confine fire from spreading to or from adjacent areas of the plant)

- provides heat transfer

- provides mechanical closure

- provides pressure-retaining boundary so that sufficient flow at adequate pressure is delivered

- provides conversion of liquid into spray

- provides flow restriction

In LRA Table 2.3.3.15, the applicant identified the following fire protection system component types within the scope of license renewal and subject to an AMR:

- closure bolting
- dikes
- expansion joint
- fire barrier penetration seals
- fire barrier walls and slabs
- fire doors
- fire hydrant
- fire rated enclosures
- flexible hose
- flow element (Annubar)
- gas bottles (CO_2, halon storage cylinders)
- gauge snubber
- gear box
- heat exchangers
- hose manifold
- odorizer
- piping and fittings
- pump casing (redundant fire pump)
- pump casing (vertical turbine)
- restricting orifice
- spray nozzle (CO_2, halon)
- sprinkler heads
- strainer
- strainer body
- tank heater
- tanks (CO_2)

- tanks (fuel oil)
- tanks (retarding chamber)
- tanks (water storage)
- thermowell
- valve body
- water motor alarm

2.3.3.15.2 Staff Evaluation

The staff reviewed LRA Section 2.3.3.15 and UFSAR Section 9.5.1 using the evaluation methodology of SER Section 2.3. The staff conducted its review in accordance with the guidance of SRP-LR Section 2.3.

In conducting its review, the staff evaluated the system functions described in the LRA and UFSAR to verify that the applicant had not omitted from the scope of license renewal any components with intended functions under 10 CFR 54.4(a). The staff then reviewed those components that the applicant had identified as within the scope of license renewal to verify that it had not omitted any passive and long-lived components subject to an AMR in accordance with the requirements of 10 CFR 54.21(a)(1).

The staff also reviewed the approved fire protection SER dated March 3, 1978, and supplements for OCGS. This report, referenced directly in the fire protection CLB, summarizes the fire protection program and commitments to 10 CFR 50.48 with the guidance of Appendix A to BTP Auxiliary and Power Conversion Systems Branch (APCSB) 9.5-1, "Guidelines for Fire Protection for Nuclear Power Plants, Docketed Prior to July 1, 1976," dated August 23, 1976. The staff then reviewed those components that the applicant had identified as within the scope of license renewal to verify that it had not omitted any passive and long-lived components subject to an AMR under 10 CFR 54.21(a)(1). The applicant provided a technical position paper which summarizes the results of the study of the fire protection program documents and the systems and structures necessary for compliance with 10 CFR 50.48.

The staff's review of LRA Section 2.3.3.15 identified areas in which additional information was necessary to complete the evaluation of the applicant's scoping and screening results. The applicant responded to the staff's RAIs as discussed below.

In RAI 2.3.3.15-1 dated January 5, 2006, the staff stated that drawing LR-JC-19479, sheet 2, shows the sprinkler system valve for sprinkler systems 17A and 17B (C-1) colored in green (i.e., within the scope of license renewal). Drawing LR-JC-19479, sheet 3, shows sprinkler systems 17A and 17B (A-6) as not within the scope of license renewal. The staff requested that the applicant verify whether sprinkler valves 17A and 17B are within the scope of license renewal in accordance with 10 CFR 54.4(a) and subject to an AMR in accordance with 10 CFR 54.21(a)(1) or, if excluded from the scope of license renewal and not subject to an AMR, justify the exclusion.

In its response dated February 3, 2006, the applicant stated that drawing LR-JC-19479, sheet 2, inadvertently identified sprinkler systems 17A and 17B as within the scope of license renewal, that these systems are not within the scope of license renewal, and that the basis for exclusion is documented in PP-07, "Systems and Structures Relied upon to Demonstrate Compliance With 10 CFR Part 50.48 - Fire protection:"

These sprinkler systems, downstream of the isolation valve V-9-913, are classified as not important to safety (NITS) on the flow diagram and in TDR-622, "ITS/NITS Classification of Suppression Systems and Fire Detection Systems." The component record list does not identify any safety-related components in the areas covered by these sprinkler systems. The OCGS fire hazards analysis report does not identify any fire safe-shutdown equipment in these areas. A fire in these areas does not significantly increase the risk of radioactive releases to the environment. These sprinkler systems are not included in the scope of license renewal.

The applicant further stated that drawing LR-JC-19479, sheet 2, will be revised to show details for sprinkler systems 17A and 17B as black and not within the scope of license renewal.

The staff finds the applicant's response to RAI 2.3.3.15-1 acceptable. The applicant explained that sprinkler systems 17A and 17B are not within the scope of license renewal and not subject to an AMR because they do not protect any safety-related components in the areas they cover. The license renewal drawing inadvertently included highlighted portions of the sprinkler system in error. The staff concludes that the components had been correctly excluded from the scope of license renewal and from AMR. Therefore, the staff's concern described in RAI 2.3.3.15-1 is resolved.

In RAI 2.3.3.15-2 dated January 5, 2006, the staff stated that in the fire protection SER dated March 3, 1978, Sections 3.1.5 and 5.9 discuss the halon 1301 system for the cable spreading room (CSR). The LRA does not list the halon 1301 system for the CSR. The staff requested that the applicant verify whether the halon 1301 system and components are within the scope of license renewal under 10 CFR 54.4(a) and subject to an AMR under 10 CFR 54.21(a)(1) or, if excluded from the scope of license renewal and not subject to an AMR, justify the exclusion.

In its response dated February 3, 2006, the applicant stated that the referenced fire protection SER includes items marked with asterisks to indicate that the staff would require additional information for them. Section 3.1.5 of the fire protection SER is marked with an asterisk and the additional information was provided to the NRC by letter dated August 31, 1979. In this letter, halon systems were proposed for the 480V switchgear room, control room panels, and A and B battery rooms. These proposed modifications were accepted by the staff, as indicated in supplement 3 to the fire protection SER dated August 25, 1980. These halon systems are shown as within the scope of license renewal on drawing LR-JC-19629, sheet 2. Halon systems are included in the fire protection system for license renewal.

The staff finds the applicant's response to RAI 2.3.3.15-2 acceptable because the CSR halon system had been replaced by the deluge sprinkler system during modifications. This replacement was confirmed by drawing LR-JC-19479, sheet 2. Also the staff confirmed that the CSR deluge sprinkler system is within the scope of license renewal under 10 CFR 54.4(a) and subject to an AMR under 10 CFR 54.21(a). Therefore, the staff's concern described in RAI 2.3.3.15-2 is resolved.

In RAI 2.3.3.15-3, dated January 5, 2006, the staff stated that in the SER dated March 3, 1978, Section 3.1.6 discusses automatic water spray and detection systems to protect safety-related cabling on the 23- and 51-foot levels of the reactor building and safety-related cables below the 4160V switchgear vault. The LRA does not list automatic spray systems for these areas. The staff requested that the applicant verify whether the automatic spray system and components are within the scope of license renewal under 10 CFR 54.4(a) and subject to an AMR under 10 CFR 54.21(a)(1) or, if excluded from the scope of license renewal and not subject to an AMR, that the applicant justify the exclusion.

In its response dated February 3, 2006, the applicant stated that the referenced SER includes items marked with asterisks to indicate that the NRC staff will require additional information for them, that Section 3.1.6 is marked with an asterisk, and that additional information was provided to the staff by letter dated August 31, 1979. In this letter, water spray systems were proposed for the 23- and 51-foot levels of the reactor building and for the CSR. These proposed modifications were accepted by the staff, as indicated in supplement 3 to the fire protection SER dated August 25, 1980. These systems identified as deluge systems 4A, 4B, 5, 6, 7 and 8 on drawing LR-JC-19479 sheet 2 (F-2) are shown as within the scope of license renewal on drawings LR-JC-19629 sheet 2 (typical details) and sheet 3 (B-4, F-5, C-5, G-5, B-5). Automatic spray systems are included in the fire protection system for license renewal.

The staff finds the applicant's response to RAI 2.3.3.15-3 acceptable because it adequately explained that the fire suppression systems in question are within the scope of license renewal under 10 CFR 54.4(a) and subject to an AMR under 10 CFR 54.21(a). Further, the applicant properly identified fire suppression as deluge systems represented in the drawings LR-JC-19479, sheet 2, and LR-JC-19629, sheet 2. Therefore, the staff's concern described in RAI 2.3.3.15-3 is resolved.

In RAI 2.3.3.15-4 dated January 5, 2006, the staff stated that the SER dated March 3, 1978, Section 3.1.7, discusses sprinkler systems for:

- the metal deck roof at the 119-foot level of the reactor building
- spent fuel pool cooling pumps
- the monitor and change room above and below the suspended ceiling to protect cables above the ceiling
- diesel-driven fire pumps and outside fuel oil storage tanks
- the turbine building above cable trays at the ceiling level of the condenser bay along the west wall.

The staff requested that the applicant verify whether the sprinkler system and components are within the scope of license renewal under 10 CFR 54.4(a) and subject to an AMR under 10 CFR 54.21(a)(1) or, if excluded from the scope of license renewal and not subject to an AMR, that the applicant justify the exclusion.

In its response dated February 3, 2006, the applicant stated that the referenced SER includes items marked with asterisks to indicate that the staff would require additional information for them. In the SER dated March 3, 1978, Section 3.1.7 is marked with an asterisk, and the additional information was provided to the staff by letter dated August 31, 1979. In this letter, sprinkler systems were proposed for the 119-foot level of the reactor building, spent fuel pool cooling pumps, the monitor and change area, the fire water pump house, diesel fuel tanks, condenser bay, and turbine building basement. These proposed modifications were accepted by the staff, as indicated in supplement 3 to the fire protection SER dated August 25, 1980. These systems are identified as sprinkler systems 1, 2, 3, 10, 11 and 12, and deluge system 9 on drawing LR-JC-19479, sheet 2 (F-2, G-2) and as within the scope of license renewal on drawing

LR-JC-19629, sheet 2 (typical details), and sheet 3 (D-5, G-7, E-4, C-4, E-9). Sprinkler systems are included in the fire protection system for license renewal.

The staff finds the applicant's response to RAI 2.3.3.15-4 acceptable because it adequately explained that the fire suppression systems in question are within the scope of license renewal under 10 CFR 54.4(a) and subject to an AMR under 10 CFR 54.21(a). Further, the applicant properly identified fire suppression as deluge systems represented in the drawings LR-JC-19479, sheet 2, and LR-JC-19629, sheet 2. Therefore, the staff's concern described in RAI 2.3.3.15-4 is resolved.

In RAI 2.3.3.15-5 dated January 5, 2006, the staff stated that in the SER dated March 3, 1978, Section 3.1.21 discusses water shields, dikes, or other protection that will be provided where breaks of suppression system piping may damage safety-related equipment. The staff requested that the applicant clarify whether these water shields had been installed and, if so, whether they are within the scope of license renewal under 10 CFR 54.4(a) and subject to an AMR under 10 CFR 54.21(a)(1) or, if excluded from the scope of license renewal and not subject to an AMR, that the applicant justify the exclusion.

In its response dated February 3, 2006, the applicant stated that the referenced SER includes items marked with asterisks to indicate that the staff would require additional information for them. In the SER dated March 3, 1978, Section 3.1.21 is marked with an asterisk, and the additional information was provided to the staff by letter dated August 31, 1979. This letter describes the specific design features to preclude fire protection system water damage to safety-related equipment. Curbs, drains and water shields were installed. These proposed modifications were accepted by the staff, as indicated in supplement 3 to the fire protection SER dated August 25, 1980. The in-scope curbs and spray shields are identified with the reactor building structure. The in-scope drains are identified as parts of the reactor building floor and equipment drains system, the miscellaneous floor and equipment drain system, and the roof drains and overboard discharge system shown on drawings LR-JC-147434, sheet 3, and LR-JC-2005, sheet 2.

The staff finds the applicant's response to RAI 2.3.3.15-5 acceptable because it adequately explained that the components in question are within the scope of license renewal in accordance with 10 CFR 54.4(a) and subject to an AMR in accordance with 10 CFR 54.21(a) and correctly identified them on drawings LR-JC-147434, sheet 3, and LR-JC-2005, sheet 2 as within the scope of license renewal and subject to an AMR. Therefore, the staff's concern described in RAI 2.3.3.15-5 is resolved.

2.3.3.15.3 Conclusion

The staff reviewed the LRA and the RAI responses to determine whether any SSCs that should be within the scope of license renewal had not been identified by the applicant. No omissions were identified. In addition, the staff determined whether any components subject to an AMR had not been identified by the applicant. No omissions were identified. The staff concludes that there is reasonable assurance that the applicant has adequately identified the fire protection system components within the scope of license renewal, as required by 10 CFR 54.4(a), and those subject to an AMR, as required by 10 CFR 54.21(a)(1).

2.3.3.16 Fuel Storage and Handling Equipment

2.3.3.16.1 Summary of Technical Information in the Application

In LRA Section 2.3.3.16, the applicant described the fuel storage and handling equipment system, the purpose of which is to support, transfer, and provide for storage of nuclear fuel in a manner that precludes inadvertent criticality. The fuel storage and handling equipment system is comprised of the spent fuel storage pool and racks, the new fuel storage vault and racks, the cask drop protection system, and fuel handling equipment. The spent fuel storage pool is enclosed and an integral part of the reactor building structure. It is a reinforced concrete structure completely lined with seam-welded stainless steel liner plate that serves as a watertight barrier. The pool contains 14 high-density stainless steel poison racks for storage of spent fuel, ten equipped with Boraflex and four with Boral poison. The pool is filled with 38 feet of demineralized water (25 feet above the fuel) for adequate shielding for normal building occupancy by operating personnel. Water temperature is maintained within acceptable limits by the spent fuel pool cooling system. The spent fuel storage pool and the racks are classified as safety-related seismic Class I structures. The new fuel storage vault is located within the reactor building adjacent to the spent fuel storage pool. The reinforced concrete vault contains aluminum racks for dry storage of new fuel bundles. The new fuel storage vault and the racks are classified as seismic Class I structures. The cask drop protection system is a cylindrical stainless steel guide structure assembly permanently installed in the northeast corner of the spent fuel storage pool. The guide structure assembly consists of an upper guide cylinder and a lower dashpot cylinder. The cask drop protection system rests on the bottom of the spent fuel pool and is laterally braced from the pool walls. The structure is classified seismic Class I. Fuel handling equipment consists of the reactor building overhead bridge crane, jib cranes, the refueling platform, fuel preparation machines, and special purpose tools for handling new fuel, spent fuel, and reactor vessel internals and components.

The fuel storage and handling equipment system contains safety-related components relied upon to remain functional during and following DBEs. The failure of nonsafety-related SSCs in the fuel storage and handling equipment system potentially could prevent the satisfactory accomplishment of a safety-related function.

The intended functions within the scope of license renewal include:

- provides neutron absorption in spent fuel pool

- provides structural support or structural integrity to preclude nonsafety-related component interactions that could prevent satisfactory accomplishment of a safety-related function

In LRA Table 2.3.3.16, the applicant identified the following fuel storage and handling equipment system component types within the scope of license renewal and subject to an AMR:

- cask drop protection cylindrical structure
- fuel grapple/mast
- fuel preparation machine
- new fuel storage racks
- refueling platform
- spent fuel storage racks

- structural bolt

2.3.3.16.2 Staff Evaluation

The staff reviewed LRA Section 2.3.3.16 and UFSAR Section 9.1 using the evaluation methodology of SER Section 2.3. The staff conducted its review in accordance with the guidance of SRP-LR Section 2.3.

In conducting its Tier-1 review of the BOP two-tier review process, the staff evaluated the system functions described in the LRA and UFSAR to verify that the applicant had not omitted from the scope of license renewal any components with intended functions under 10 CFR 54.4(a). The staff then reviewed those components that the applicant had identified as within the scope of license renewal to verify that it had not omitted any passive and long-lived components subject to an AMR in accordance with the requirements of 10 CFR 54.21(a)(1).

2.3.3.16.3 Conclusion

The staff reviewed the LRA to determine whether any SSCs that should be within the scope of license renewal had not been identified by the applicant. No omissions were identified. In addition, the staff determined whether any components subject to an AMR had not been identified by the applicant. No omissions were identified. The staff concludes that there is reasonable assurance that the applicant has adequately identified the fuel storage and handling equipment system components within the scope of license renewal, as required by 10 CFR 54.4(a), and those subject to an AMR, as required by 10 CFR 54.21(a)(1).

2.3.3.17 Hardened Vent System

2.3.3.17.1 Summary of Technical Information in the Application

In LRA Section 2.3.3.17, the applicant described the hardened vent system (HVS), the purpose of which is to vent the primary containment via the torus (primary path) or drywell (secondary path) during severe accident sequences that involve loss of decay heat removal capability (the torus is the preferred vent path because of the scrubbing effect of the torus water). The HVS accomplishes this purpose by providing a vent path to the ventilation stack from either the torus or drywell through the CIS nitrogen purge header and its drywell and torus nitrogen purge inlet pressure control valves. The HVS is designed for the mitigation of severe accident sequences beyond the DBA.

The failure of nonsafety-related SSCs in the HVS potentially could prevent the satisfactory accomplishment of a safety-related function.

The intended functions within the scope of license renewal include:

- provides mechanical closure
- provides pressure-retaining boundary; fission product barrier; containment isolation for fission product retention; or containment, holdup, and plateout function

In LRA Table 2.3.3.17, the applicant identified the following HVS component types within the scope of license renewal and subject to an AMR:

- closure bolting
- enclosure boot
- piping and fittings
- valve body

2.3.3.17.2 Staff Evaluation

The staff reviewed LRA Section 2.3.3.17 and UFSAR Section 6.2.7 using the evaluation methodology of SER Section 2.3. The staff conducted its review in accordance with the guidance of SRP-LR Section 2.3.

The staff reviewed the system functions described in the LRA and UFSAR to verify that the applicant had not omitted from the scope of license renewal any components with intended functions under 10 CFR 54.4(a). The staff then reviewed components that the applicant had identified as within the scope of license renewal to verify that it had not omitted any passive and long-lived components subject to an AMR in accordance with the requirements of 10 CFR 54.21(a)(1).

2.3.3.17.3 Conclusion

The staff reviewed the LRA to determine whether any SSCs that should be within the scope of license renewal had not been identified by the applicant. No omissions were identified. In addition, the staff determined whether any components subject to an AMR had not been identified by the applicant. No omissions were identified. The staff concludes that there is reasonable assurance that the applicant has adequately identified the HVS components within the scope of license renewal, as required by 10 CFR 54.4(a), and those subject to an AMR, as required by 10 CFR 54.21(a)(1).

2.3.3.18 Heating & Process Steam System

2.3.3.18.1 Summary of Technical Information in the Application

In LRA Section 2.3.3.18, the applicant described the heating and process steam system, the purpose of which is to provide steam in sufficient capacity for operation of the radwaste concentrator for evaporative processing of liquid radioactive waste, for plant area heating, and for oxygen-free boiler feedwater. It accomplishes its purpose through two fuel oil-fired boilers and their supporting systems, including steam distribution and condensate systems, and through chemical addition. Operation of the heating and process steam system is not required to perform or support any safety-related function and consequently the system is nonsafety-related.

The failure of nonsafety-related SSCs in the heating and process steam system potentially could prevent the satisfactory accomplishment of a safety-related function.

The intended functions within the scope of license renewal include:

- maintains mechanical and structural integrity to prevent spatial interactions that could cause failure of safety-related SSCs (includes the required structural support when the nonsafety-related leakage boundary piping is also attached to safety-related piping)
- provides mechanical closure

In LRA Table 2.3.3.18, the applicant identified the following heating and process steam system component types within the scope of license renewal and subject to an AMR:

- closure bolting
- coolers (sample)
- flexible connection
- flow element
- heat exchangers
- piping and fittings
- pump casing - chemical addition pump CH-P-11
- pump casing - condensate return pumps P-13-1A/B, chemical feed addition pumps CH-P-6A/B, boiler No. 1 feed pumps CHP-4A/B, boiler No. 2 feed pumps CH-P-3A/B, deaerator feed pumps CH-P-5A/B, chemical recirculation pump CH-P-10
- restricting orifice
- sight glasses
- soot blowers
- steam trap
- strainer body
- tanks - chemical feed addition tanks CHT-3A/B
- tanks - deaerator CH-T-2, condensate return unit T-13-1, heating boiler condensate storage tank T-13-2, heating boiler flash tank T-13-3
- valve body

2.3.3.18.2 Staff Evaluation

The staff reviewed LRA Section 2.3.3.18 and UFSAR Section 10.4.8 using the evaluation methodology of SER Section 2.3. The staff conducted its review in accordance with the guidance of SRP-LR Section 2.3.

In conducting its Tier-1 review of the BOP two-tier review process, the staff evaluated the system functions described in the LRA and UFSAR to verify that the applicant had not omitted from the scope of license renewal any components with intended functions under 10 CFR 54.4(a). The staff then reviewed those components that the applicant had identified as within the scope of license renewal to verify that it had not omitted any passive and long-lived components subject to an AMR in accordance with the requirements of 10 CFR 54.21(a)(1).

2.3.3.18.3 Conclusion

The staff reviewed the LRA to determine whether any SSCs that should be within the scope of license renewal had not been identified by the applicant. No omissions were identified. In addition, the staff determined whether any components subject to an AMR had not been identified by the applicant. No omissions were identified. The staff concludes that there is reasonable assurance that the applicant has adequately identified the heating and process steam system components within the scope of license renewal, as required by 10 CFR 54.4(a), and those subject to an AMR, as required by 10 CFR 54.21(a)(1).

2.3.3.19 Hydrogen & Oxygen Monitoring System

2.3.3.19.1 Summary of Technical Information in the Application

In LRA Section 2.3.3.19, the applicant described the hydrogen and oxygen monitoring system, which consists of the drywell hydrogen/oxygen monitoring subsystem and the drywell and torus oxygen monitoring subsystem. The purpose of the hydrogen and oxygen monitoring system is to monitor the primary containment atmosphere to ensure that oxygen and hydrogen levels do not approach flammability limits. The hydrogen and oxygen monitoring system accomplishes this purpose post-accident and during normal power operations. During post-accident operation the drywell hydrogen/oxygen monitoring subsystem processes a drywell atmosphere sample through one of two redundant hydrogen and oxygen measuring loops. During normal power operation the drywell hydrogen/oxygen monitoring subsystem is in the standby mode except for calibration or maintenance and the drywell and torus oxygen monitoring subsystem is in service to monitor the oxygen concentration of the atmosphere in the drywell and torus areas.

The hydrogen and oxygen monitoring system contains safety-related components relied upon to remain functional during and following DBEs. The failure of nonsafety-related SSCs in the hydrogen and oxygen monitoring system potentially could prevent the satisfactory accomplishment of a safety-related function. In addition, the hydrogen and oxygen monitoring system performs functions that support EQ.

The intended functions within the scope of license renewal include:

- maintains mechanical and structural integrity to prevent spatial interactions that could cause failure of safety-related SSCs (includes the required structural support when the nonsafety-related leakage boundary piping is also attached to safety-related piping)
- provides mechanical closure
- provides pressure-retaining boundary; fission product barrier; containment isolation; or containment, holdup, and plateout (main steam system)
- provides structural support or structural integrity to preclude nonsafety-related component interactions that could prevent satisfactory accomplishment of a safety-related function
- provides flow restriction

In LRA Table 2.3.3.19, the applicant identified the following hydrogen and oxygen monitoring system component types within the scope of license renewal and subject to an AMR:

- closure bolting
- drain trap (O_2 analyzers)
- filter housing (O_2 analyzers)
- flexible hose
- flow element
- heat exchangers (air cooled)
- moisture separator (H_2O_2 analyzers)
- piping and fittings
- pump casing
- restricting orifice
- sensor element
- tanks (volume chamber)
- valve body
- water separator (O_2 analyzers)

2.3.3.19.2 Staff Evaluation

The staff reviewed LRA Section 2.3.3.19 and UFSAR Sections 6.2.5 and 7.6.1 using the evaluation methodology of SER Section 2.3. The staff conducted its review in accordance with the guidance of SRP-LR Section 2.3.

The staff reviewed the system functions described in the LRA and UFSAR to verify that the applicant had not omitted from the scope of license renewal any components with intended functions under 10 CFR 54.4(a). The staff then reviewed components that the applicant had identified as within the scope of license renewal to verify that it had not omitted any passive and long-lived components subject to an AMR in accordance with the requirements of 10 CFR 54.21(a)(1).

2.3.3.19.3 Conclusion

The staff reviewed the LRA to determine whether any SSCs that should be within the scope of license renewal had not been identified by the applicant. No omissions were identified. In addition, the staff determined whether any components subject to an AMR had not been identified by the applicant. No omissions were identified. The staff concludes that there is reasonable assurance that the applicant has adequately identified the hydrogen and oxygen monitoring system components within the scope of license renewal, as required by 10 CFR 54.4(a), and those subject to an AMR, as required by 10 CFR 54.21(a)(1).

2.3.3.20 Instrument (Control) Air System

2.3.3.20.1 Summary of Technical Information in the Application

In LRA Section 2.3.3.20, the applicant described the instrument air system, the purpose of which is to provide clean and dried compressed air to pneumatically-operated instruments and valves. To accomplish this purpose, the instrument air system receives compressed air from the service air system and processes it through air dryers for distribution to components in support of plant operation. The instrument air system also penetrates the drywell and is isolated by the closing of the instrument air containment isolation valve. This instrument air supply to the drywell is charged with nitrogen during power operation to reduce combustible gas in the drywell and torus with compressed air as a backup. During normal plant operation the service air compressors operate continuously to supply the source of the plant's required instrument and control air and keep the accumulators charged. Where required, pneumatically-operated devices are designed to fail-safe upon loss of air or are provided with accumulators to provide a stored volume of compressed air when the compressors or other nonsafety-related sections of the instrument air system are unavailable. Accumulators are isolated by check valves to ensure backup air for components credited to function during or following DBEs.

The instrument air system contains safety-related components relied upon to remain functional during and following DBEs. The failure of nonsafety-related SSCs in the instrument air system potentially could prevent the satisfactory accomplishment of a safety-related function. In addition, the instrument air system performs functions that support fire protection, SBO, and EQ.

The intended functions within the scope of license renewal include:

- provides mechanical closure
- provides pressure-retaining boundary or containment isolation
- provides structural support or structural integrity to preclude nonsafety-related component interactions that could prevent satisfactory accomplishment of a safety-related function

In LRA Table 2.3.3.20, the applicant identified the following instrument air system component types within the scope of license renewal and subject to an AMR:

- accumulator
- closure bolting
- filter housing
- flexible hose
- flow element
- piping and fittings
- valve body

2.3.3.20.2 Staff Evaluation

The staff reviewed LRA Section 2.3.3.20 and UFSAR Section 9.3.1 using the evaluation methodology of SER Section 2.3. The staff conducted its review in accordance with the guidance of SRP-LR Section 2.3.

In conducting its Tier-2 review of the BOP two-tier review process, the staff evaluated the system functions described in the LRA and UFSAR to verify that the applicant had not omitted from the scope of license renewal any components with intended functions under 10 CFR 54.4(a). The staff then reviewed those components that the applicant had identified as within the scope of license renewal to verify that it had not omitted any passive and long-lived components subject to an AMR in accordance with the requirements of 10 CFR 54.21(a)(1).

2-85

2.3.3.20.3 Conclusion

The staff reviewed the LRA to determine whether any SSCs that should be within the scope of license renewal had not been identified by the applicant. No omissions were identified. In addition, the staff determined whether any components subject to an AMR had not been identified by the applicant. No omissions were identified. The staff concludes that there is reasonable assurance that the applicant has adequately identified the instrument air system components within the scope of license renewal, as required by 10 CFR 54.4(a), and those subject to an AMR, as required by 10 CFR 54.21(a)(1).

2.3.3.21 Main Fuel Oil Storage & Transfer System

2.3.3.21.1 Summary of Technical Information in the Application

In LRA Section 2.3.3.21, the applicant described the main fuel oil storage and transfer system, a mechanical system designed to store and transfer fuel oil to the heating and process steam system and to the emergency diesel generator fuel storage tank under normal plant operating conditions. The main fuel oil storage and transfer system receives fuel oil from tank trucks and stores it in a tank located in the yard. Fuel oil is conveyed to the Nos. 1 and 2 heating boilers by a transfer pump, pressurized by boiler fuel pumps, and fed to the boilers for combustion. The system supplies bottled propane to both heating boilers for ignition and atomizing air to the No. 2 heating boiler. The system can be aligned to provide fuel oil to the EDG fuel oil tank but is not credited for diesel generator operation.

The failure of nonsafety-related SSCs in the main fuel oil storage and transfer system potentially could prevent the satisfactory accomplishment of a safety-related function.

The intended functions within the scope of license renewal include:

- maintains mechanical and structural integrity to prevent spatial interactions that could cause failure of safety-related SSCs (includes the required structural support when the nonsafety-related leakage boundary piping is also attached to safety-related piping)
- provides mechanical closure

In LRA Table 2.3.3.21, the applicant identified the following main fuel oil storage and transfer system component types within the scope of license renewal and subject to an AMR:

- closure bolting
- flexible hose
- flow meter
- piping and fittings
- pump casing
- sight glasses
- strainer body
- valve body

2.3.3.21.2 Staff Evaluation

The staff reviewed LRA Section 2.3.3.21 and UFSAR Sections 9.5.4 and 10.4.8 using the evaluation methodology of SER Section 2.3. The staff conducted its review in accordance with the guidance of SRP-LR Section 2.3.

In conducting its Tier-1 review of the BOP two-tier review process, the staff evaluated the system functions described in the LRA and UFSAR to verify that the applicant had not omitted from the scope of license renewal any components with intended functions under 10 CFR 54.4(a). The staff then reviewed those components that the applicant had identified as within the scope of license renewal to verify that it had not omitted any passive and long-lived components subject to an AMR in accordance with the requirements of 10 CFR 54.21(a)(1).

2.3.3.21.3 Conclusion

The staff reviewed the LRA to determine whether any SSCs that should be within the scope of license renewal had not been identified by the applicant. No omissions were identified. In addition, the staff determined whether any components subject to an AMR had not been identified by the applicant. No omissions were identified. The staff concludes that there is reasonable assurance that the applicant has adequately identified the main fuel oil storage and transfer system components within the scope of license renewal, as required by 10 CFR 54.4(a), and those subject to an AMR, as required by 10 CFR 54.21(a)(1).

2.3.3.22 Miscellaneous Floor and Equipment Drain System

2.3.3.22.1 Summary of Technical Information in the Application

In LRA Section 2.3.3.22, the applicant described the miscellaneous floor and equipment drain (MFED) system, the purpose of which is to collect floor drains and equipment drains in various locations throughout the site and transfer the collected drainage to the radwaste system for processing, overboard discharge, or disposal. The MFED system accomplishes this purpose though use of gravity drain lines, sumps, tanks, pumps, and monitoring instruments used to collect and classify waste drainage. The MFED system is designed to accommodate the volumes of fluids from maintenance activities, system flushing, rinsing operations, and other plant work and is sized to minimize any potential for plant flooding. Floor drains in the cable spreading rooms of the turbine building are credited in existing analyses with accommodating water flow from actuation of the fire suppression systems in those rooms. The MFED system consists of turbine building floor and equipment drains, offgas building floor and equipment drains, radwaste floor and equipment drains, laundry and laboratory drains, miscellaneous building sumps, condensate transfer building sump, and miscellaneous oil drain systems.

The failure of nonsafety-related SSCs in the MFED system potentially could prevent the satisfactory accomplishment of a safety-related function. The MFED system also performs functions that support fire protection.

The intended functions within the scope of license renewal include:

- maintains mechanical and structural integrity to prevent spatial interactions that could cause failure of safety-related SSCs (includes the required structural support when the nonsafety-related leakage boundary piping is also attached to safety-related piping)

- provides mechanical closure
- provides pressure-retaining boundary

In LRA Table 2.3.3.22, the applicant identified the following MFED system component types within the scope of license renewal and subject to an AMR:

- closure bolting
- flexible hose
- piping and fittings
- pump casing (lab drain tank pump P-22-003)
- pump casing (laundry drain tank pump P-22-002)
- pump casings (regeneration waste transfer pumps P-22-28A,B and P-22-29A,B)
- strainer body
- tanks (lab drain tank T-22-003)
- tanks (laundry drain tank T-22-002)
- tanks (oil separator DS-Y-105 and oil receiver DS-T-1)
- tanks (regeneration system waste tank 1-1 low and high conductivity compartments)
- valve body

2.3.3.22.2 Staff Evaluation

The staff reviewed LRA Section 2.3.3.22 and UFSAR Sections 9.3.3 and 11.2.2 using the evaluation methodology of SER Section 2.3. The staff conducted its review in accordance with the guidance of SRP-LR Section 2.3.

In conducting its Tier-1 review of the BOP two-tier review process, the staff evaluated the system functions described in the LRA and UFSAR to verify that the applicant had not omitted from the scope of license renewal any components with intended functions under 10 CFR 54.4(a). The staff then reviewed those components that the applicant had identified as within the scope of license renewal to verify that it had not omitted any passive and long-lived components subject to an AMR in accordance with the requirements of 10 CFR 54.21(a)(1).

LRA Section 2.3.3.22 states that the heating boiler house contains some liquid-filled portions of the MFED system in proximity to equipment performing a safety-related function and thus within the scope of license renewal because they perform a 10 CFR 54.4(a)(2) function. LRA Section 2.4.10 states that the old heating boiler house contains several safety-related electrical components and that major components housed in the buildings (old and new heating boiler houses) include oil-fired boilers, heating boiler feed pumps, fuel oil pumps, deaerator, chemical tanks and feed pumps, boiler condensate storage tank, and system piping. The staff determined that there was insufficient information to determine which MFED system component types in the old heating boiler house are within the scope of license renewal. The staff referred this issue to NRC Region I for review to verify which MFED system components, if any, located in proximity to the safety-related components in the old heating boiler house are within the scope of license renewal for the purposes of 10 CFR 54.4(a)(2).

Subsequently, the NRC resident inspector reviewed the applicant's piping and instrumentation drawings and did a system walkdown of the MFED equipment located in the old heating boiler house. The resident inspector concludes that the only safety-related equipment in the boiler house is the motor control center for the standby gas treatment system exhaust fans. The remaining equipment is boiler-related or diesel fuel oil transfer from the storage tank not safety-related or credited in the design bases. The resident inspector verified that the MFED equipment including the piping, fittings, valves and oil separator is located within the old heating boiler house near the safety-related standby gas treatment motor control center. Based on this information the staff concludes that the MFED equipment shown on drawing LR-JC-147434 sheet 2, is correctly identified as within the scope of license renewal. The staff's concern is resolved.

2.3.3.22.3 Conclusion

The staff reviewed the LRA to determine whether any SSCs that should be within the scope of license renewal had not been identified by the applicant. No omissions were identified. In addition, the staff determined whether any components subject to an AMR had not been identified by the applicant. No omissions were identified. The staff concludes that there is reasonable assurance that the applicant has adequately identified the MFED system components within the scope of license renewal, as required by 10 CFR 54.4(a), and those subject to an AMR, as required by 10 CFR 54.21(a)(1).

2.3.3.23 Nitrogen Supply System

2.3.3.23.1 Summary of Technical Information in the Application

In LRA Section 2.3.3.23, the applicant described the nitrogen supply system, the purpose of which is to supply vaporized nitrogen at a specified pressure and temperature to the CIS, drywell nitrogen subsystem, traveling in-core probe (TIP) system indexing mechanisms, and feedwater heaters. The nitrogen supply system accomplishes this purpose by processing stored liquid nitrogen through a vaporizer, heaters, and pressure regulating valves and providing it to the CIS, drywell nitrogen sub-system, TIP system indexing mechanisms, and feedwater heaters on demand. The nitrogen supply system also provides nitrogen to the reactor water cleanup (RWCU) system recirculation pump surge tank and the CRD system accumulator nitrogen charging system. This portion of the nitrogen supply system consists of local bottled nitrogen supplies, pressure regulators, and piping. The nitrogen supply system is manually initiated to support its users. The nitrogen supply to the TIP system indexing mechanisms penetrates the primary containment and is provided with containment isolation devices.

The nitrogen supply system contains safety-related components relied upon to remain functional during and following DBEs. In addition, the nitrogen supply system performs functions that support fire protection.

The intended functions within the scope of license renewal include:

- provides filtration
- provides heat transfer
- provides mechanical closure
- provides pressure-retaining boundary or containment isolation
- provides flow restriction

In LRA Table 2.3.3.23, the applicant identified the following nitrogen supply system component types within the scope of license renewal and subject to an AMR:

- closure bolting
- drip leg
- heat exchangers (electric heater)
- heat exchangers (trim heater)
- heat exchangers (vaporizer)
- piping and fittings
- pressure building coils
- restricting orifice
- rupture disks
- sight glasses (flow indication)
- strainer
- strainer body
- tanks
- thermowell
- valve body

2.3.3.23.2 Staff Evaluation

The staff reviewed LRA Section 2.3.3.23 and UFSAR Sections 1.9.21, 3.1.37, 6.2.5, and Table 6.2-12 using the evaluation methodology of SER Section 2.3. The staff conducted its review in accordance with the guidance of SRP-LR Section 2.3.

In conducting its Tier-2 review of the BOP two-tier review process, the staff evaluated the system functions described in the LRA and UFSAR to verify that the applicant had not omitted from the scope of license renewal any·components with intended functions under 10 CFR 54.4(a). The staff then reviewed those components that the applicant had identified as within the scope of license renewal to verify that it had not omitted any passive and long-lived components subject to an AMR in accordance with the requirements of 10 CFR 54.21(a)(1).

The staff of LRA Section 2.3.3.23 identified an area in which additional information was necessary to complete the evaluation of the applicant's scoping and screening results. The applicant responded to the staff's RAI as discussed below.

In RAI 2.3.3.23-1 dated December 28, 2005, the staff stated that although the LRA Section 2.3.3.23 drawing shows a 3/8-inch nitrogen supply line to the neutron monitoring system penetrating primary containment to have an intended function outside containment it has no intended function inside containment. No explanation is given for the change in intended function for the nitrogen line. Therefore, the staff requested that the applicant confirm whether the nitrogen line has no intended function inside containment as the neutron monitoring system has components within the scope of license renewal.

In its response dated January 26, 2006, the applicant stated:

> The 3/8" line penetrating the drywell at penetration X-45 as shown on license renewal drawing LR-SN-13432.19-1, drawing coordinate A-3, is the "TIP purge instrumentation reference leg piping" as described in the system boundary discussion of LRA Section 2.3.3.23 for the nitrogen supply system.

LRA Section 2.3.3.23 states, "The Nitrogen Supply System supports the primary containment boundary intended function. This portion of the system includes the nitrogen supply to the TIP System indexers starting from the automatic containment isolation valve and continuing to the containment penetration. Also included is the TIP purge instrumentation reference leg piping from the containment penetration up to and including the manual isolation valve." Inboard of the TIP purge and TIP purge instrumentation reference leg piping containment isolation valves is also discussed in the system boundary discussion of LRA Section 2.3.3.23.

As stated in LRA Section 2.3.3.23, the nitrogen supply lines up to these valves are included in scope as they define the nitrogen supply system pressure boundary necessary to support the intended function for fire protection.

The nitrogen piping inside the primary containment associated with the TIP system is not required to functionally support the intended functions of the neutron monitoring system (NMS). Furthermore, as stated in LRA Section 2.1.5.2, nonsafety-related systems containing air or gas are not included in the scope of license renewal for 10 CFR 54.4(a)(2) spatial interaction.

Therefore, AmerGen has concluded that the 3/8" nitrogen supply to the NMS is not within the scope of license renewal. Additionally, the supports for the nitrogen supply system piping inside of the primary containment are included in scope to prevent the piping from falling and potentially impacting safety-related SSCs. These supports are evaluated on a commodity level and are not included in the evaluation of the nitrogen supply system.

The staff review finds the applicant's response acceptable because the TIP system nitrogen piping inside the primary containment is nonsafety-related and does not functionally support the intended functions of the NMS. As such, the piping in question satisfies none of the 10 CFR 54.4(a) scoping criteria. Therefore, the staff's concern described in RAI 2.3.3.23-1 is resolved.

2.3.3.23.3 Conclusion

The staff reviewed the LRA and the RAI response to determine whether any SSCs that should be within the scope of license renewal had not been identified by the applicant. No omissions were identified. In addition, the staff determined whether any components subject to an AMR had not been identified by the applicant. No omissions were identified. The staff concludes that there is reasonable assurance that the applicant has adequately identified the nitrogen supply system components within the scope of license renewal, as required by 10 CFR 54.4(a), and those subject to an AMR, as required by 10 CFR 54.21(a)(1).

2.3.3.24 Noble Metals Monitoring System

2.3.3.24.1 Summary of Technical Information in the Application

In LRA Section 2.3.3.24, the applicant described the noble metals monitoring system (NMMS), a reactor coolant monitoring system designed for determining the effectiveness of the noble metal chemical addition injection process performed during the 1R19 refueling outage. The purpose of the NMMS is to track and trend the integrity of the noble metals film applied to the reactor internals and recirculation piping to ensure its ability to support hydrogen water chemistry (HWC) in the mitigation of intergranular stress corrosion cracking (IGSCC). The NMMS accomplishes this purpose by monitoring the electrochemical corrosion potential of the reactor coolant, simulating and trending noble metals deposition, and monitoring and recording NMMS parameters. Manual valves local to the NMMS are used to place the system in service. The NMMS is operated when the plant is at power and the RWCU system is in operation.

The failure of nonsafety-related SSCs in the NMMS potentially could prevent the satisfactory accomplishment of a safety-related function.

The intended functions within the scope of license renewal include:

- maintains mechanical and structural integrity to prevent spatial interactions that could cause failure of safety-related SSCs (includes the required structural support when the nonsafety-related leakage boundary piping is also attached to safety-related piping)
- provides mechanical closure

In LRA Table 2.3.3.24, the applicant identified the following NMMS component types within the scope of license renewal and subject to an AMR:

- closure bolting
- flow element
- piping and fittings
- sensor element
- valve body

2.3.3.24.2 Staff Evaluation

The staff reviewed LRA Section 2.3.3.24 and UFSAR Section 5.2.3.4 using the evaluation methodology of SER Section 2.3. The staff conducted its review in accordance with the guidance of SRP-LR Section 2.3.

In conducting its review, the staff evaluated the system functions described in the LRA and UFSAR to verify that the applicant had not omitted from the scope of license renewal any components with intended functions under 10 CFR 54.4(a). The staff then reviewed those components that the applicant had identified as within the scope of license renewal to verify that it had not omitted any passive and long-lived components subject to an AMR in accordance with the requirements of 10 CFR 54.21(a)(1).

2.3.3.24.3 Conclusion

The staff reviewed the LRA to determine whether any SSCs that should be within the scope of license renewal had not been identified by the applicant. No omissions were identified. In addition, the staff determined whether any components subject to an AMR had not been identified by the applicant. No omissions were identified. The staff concludes that there is reasonable assurance that the applicant has adequately identified the NMMS components within the scope of license renewal, as required by 10 CFR 54.4(a), and those subject to an AMR, as required by 10 CFR 54.21(a)(1).

2.3.3.25 Post-Accident Sampling System

2.3.3.25.1 Summary of Technical Information in the Application

In LRA Section 2.3.3.25, the applicant described the post-accident sampling system (PASS) designed to obtain liquid and gaseous samples from the primary containment, gaseous samples from the secondary containment, and liquid samples from the reactor vessel for radiological and chemical analysis to estimate post-accident core damage and coolant corrosiveness. Reactor coolant samples can be drawn from reactor recirculation Loop A, the liquid poison system piping, and the SCS piping. A torus water sample can be drawn from the core spray system piping. The samples pass through sample coolers located in the reactor building TIP room and continue to the sample station in the PASS room. All liquid samples are returned to the primary containment through the core spray pumps suction line during accident conditions. Gaseous atmosphere samples can be obtained from the drywell and wetwell through the hydrogen and oxygen monitoring system. A secondary containment atmosphere sample also can be drawn into the PASS station. Primary containment gas samples are returned to the drywell, and secondary containment gas samples are returned to the reactor building atmosphere. The PASS was originally installed as required by the NRC and as described in NUREG-0737. While no longer required by the technical specifications, the PASS continues to be maintained and operation of the system is described in approved plant procedures.

The PASS contains safety-related components relied upon to remain functional during and following DBEs. The failure of nonsafety-related SSCs in the PASS potentially could prevent the satisfactory accomplishment of a safety-related function. In addition, the PASS performs functions that support EQ.

The intended functions within the scope of license renewal include:

- maintains mechanical and structural integrity to prevent spatial interactions that could cause failure of safety-related SSCs (includes the required structural support when the nonsafety-related leakage boundary piping is also attached to safety-related piping)
- provides mechanical closure
- provides pressure-retaining boundary; fission product barrier; or containment isolation
- provides structural support or structural integrity to preclude nonsafety-related component interactions that could prevent satisfactory accomplishment of a safety-related function

In LRA Table 2.3.3.25, the applicant identified the following PASS component types within the scope of license renewal and subject to an AMR:

- closure bolting
- piping and fittings
- valve body

2.3.3.25.2 Staff Evaluation

The staff reviewed LRA Section 2.3.3.25 and UFSAR Sections 1.9 and 11.5.2.12 using the evaluation methodology of SER Section 2.3. The staff conducted its review in accordance with the guidance of SRP-LR Section 2.3.

In conducting its Tier-2 review of the BOP two-tier review process, the staff evaluated the system functions described in the LRA and UFSAR to verify that the applicant had not omitted from the scope of license renewal any components with intended functions under 10 CFR 54.4(a). The staff then reviewed those components that the applicant had identified as within the scope of license renewal to verify that it had not omitted any passive and long-lived components subject to an AMR in accordance with the requirements of 10 CFR 54.21(a)(1).

2.3.3.25.3 Conclusion

The staff reviewed the LRA to determine whether any SSCs that should be within the scope of license renewal had not been identified by the applicant. No omissions were identified. In addition, the staff determined whether any components subject to an AMR had not been identified by the applicant. No omissions were identified. The staff concludes that there is reasonable assurance that the applicant has adequately identified the PASS components within the scope of license renewal, as required by 10 CFR 54.4(a), and those subject to an AMR, as required by 10 CFR 54.21(a)(1).

2.3.3.26 Process Sampling System

2.3.3.26.1 Summary of Technical Information in the Application

In LRA Section 2.3.3.26, the applicant described the process sampling system designed to permit a representative sample to be taken in a form which can be used in the laboratory and which safeguards against change in the constituents to be examined, minimizes the contamination and radiation at the sample point, and reduces decay and sample line plateout as much as possible. The purpose of the process sampling system is to monitor the operation of equipment and to supply information for making operating decisions where these are influenced by water chemistry. It accomplishes this purpose by collecting steam, gaseous, and liquid samples throughout the facility. Sample stream flow rates are selected to maintain turbulent flow for more accurate sampling. The process sampling system is comprised of the following subsystems: reactor sampling subsystem, radwaste sampling subsystem, composite sample subsystem, hydrogen detection/sampling subsystem, and the off-gas sample subsystem. The reactor sampling subsystem consists of the reactor water sample station and the final feedwater facility. The reactor water sample station provides sample and analysis capabilities for reactor water and the RWCU system. The final feedwater facility system consists of sampling of the turbine building primary systems. The radwaste sampling system monitors activity at various points of the radwaste system, which is a liquid and solid radioactive waste management system. In the composite sample subsystem, composite samples of condenser cooling water are taken locally at the plant's intake and outfall. The hydrogen detection/sampling subsystem monitors the augmented off-gas recombiner subsystem. The off-gas sample subsystem takes a sample at the air ejectors to measure activity release and H_2O_2 and air leakage, a sample at the stack to measure particulate and iodine release, and a sample at the inlet and outlet of the offgas filter to determine filter efficiency.

The failure of nonsafety-related SSCs in the process sampling system potentially could prevent the satisfactory accomplishment of a safety-related function.

The intended functions within the scope of license renewal include:

- maintains mechanical and structural integrity to prevent spatial interactions that could cause failure of safety-related SSCs (includes the required structural support when the nonsafety-related leakage boundary piping is also attached to safety-related piping)
- provides mechanical closure

In LRA Table 2.3.3.26, the applicant identified the following process sampling system component types within the scope of license renewal and subject to an AMR:

- closure bolting
- coolers
- evaporator
- flexible hose
- flow element
- piping and fittings
- pump casing
- sensor element
- sight glasses
- tanks (reservoir)
- thermowell
- valve body

2.3.3.26.2 Staff Evaluation

The staff reviewed LRA Section 2.3.3.26 and UFSAR Section 9.3.2 and Table 9.3-3 using the evaluation methodology of SER Section 2.3. The staff conducted its review in accordance with the guidance of SRP-LR Section 2.3.

In conducting its Tier-1 review of the BOP two-tier review process, the staff evaluated the system functions described in the LRA and UFSAR to verify that the applicant had not omitted from the scope of license renewal any components with intended functions under 10 CFR 54.4(a). The staff then reviewed those components that the applicant had identified as within the scope of license renewal to verify that it had not omitted any passive and long-lived components subject to an AMR in accordance with the requirements of 10 CFR 54.21(a)(1).

The staff's review of LRA Section 2.3.3.26 identified an area in which additional information was necessary to complete the evaluation of the applicant's scoping and screening results. The applicant responded to the staff's RAI as discussed below.

In RAI 2.3.3.26-1 dated December 28, 2005, the staff stated that on drawing LR-GU-3E-551-21-1000 the feedwater sample sink and the condensate sample sink are shown within the scope of license renewal; however, "sinks" are not listed as components subject to an AMR. Therefore, the staff requested that the applicant indicate whether the sinks are included within a component type subject to an AMR or justify their exclusion from an AMR.

In its response dated January 26, 2006, the applicant stated:

> The feedwater and condensate sample sinks are correctly shown on license renewal drawing LR-GU-3E-551-21-1000 as in scope for spatial interaction (10 CFR 54.4(a)(2)). LRA Table 2.3.3.26 for process sampling system components subject to aging management review and LRA Table 3.3.2.1.26 for process sampling system aging management evaluation should have included a component type of "sinks," or equivalently named component, with an intended function of "leakage boundary."

> Attachment I to this enclosure identifies the addition of "sinks" to Tables 2.3.3.26 and 3.3.2.1.26.

The staff review finds the applicant's response acceptable because it appropriately added "sinks" as a component type subject to an AMR in accordance with 10 CFR 54.21(a)(1) and identified the component intended function. Therefore, the staff's concern described in RAI 2.3.3.26-1 is resolved.

2.3.3.26.3 Conclusion

The staff reviewed the LRA and the RAI response to determine whether any SSCs that should be within the scope of license renewal had not been identified by the applicant. No omissions were identified. In addition, the staff determined whether any components subject to an AMR had not been identified by the applicant. No omissions were identified. The staff concludes that there is reasonable assurance that the applicant has adequately identified the process sampling system components within the scope of license renewal, as required by 10 CFR 54.4(a), and those subject to an AMR, as required by 10 CFR 54.21(a)(1).

2.3.3.27 Radiation Monitoring System

2.3.3.27.1 Summary of Technical Information in the Application

In LRA Section 2.3.3.27, the applicant described the radiation monitoring system, the purpose of which is to detect the release of radioactivity, monitor radiation levels in key locations throughout the plant, and monitor radioactivity concentration levels of major process system discharge streams. The system accomplishes its purpose by utilizing radiation detectors and circuitry to monitor and indicate radiation levels. The radiation monitoring system consists of process and effluent radiological monitoring, area radiation and airborne radioactivity monitoring, and containment atmosphere particulate and gaseous radioactivity monitoring. The process and effluent radiological monitoring system is designed to detect radioactive gaseous and liquid leakage, provide warning and automatic control as appropriate when radioactivity in a process stream reaches a preset limit, provide information on fuel and radioactive processing equipment performance, provide a record of radioactivity present in various plant systems, and provide a record of radioactivity released to the environment for compliance with regulatory limits. The area radiation and airborne radioactivity monitoring system is designed to monitor the level of radiation in areas where personnel access may be required, assist in maintaining occupational radiation exposures as low as reasonably achievable, alarm when radiation levels exceed preset limits, and provide a continuous record of radiation levels in key locations throughout the plant. The containment atmosphere particulate and gaseous radioactivity monitoring system provides a diverse means of RCS leak detection by detecting the release of radioactivity from a leak and subsequent flashing to steam. The system is designed to detect both particulate and noble gas radiation.

The radiation monitoring system contains safety-related components relied upon to remain functional during and following DBEs. The failure of nonsafety-related SSCs in the radiation monitoring system potentially could prevent the satisfactory accomplishment of a safety-related function.

The intended functions within the scope of license renewal include:

- provides mechanical closure
- provides pressure-retaining boundary; fission product barrier; or containment isolation
- provides structural support or structural integrity to preclude nonsafety-related component interactions that could prevent satisfactory accomplishment of a safety-related function

In LRA Table 2.3.3.27, the applicant identified the following radiation monitoring system component types within the scope of license renewal and subject to an AMR:

- closure bolting
- piping and fittings
- valve body

2.3.3.27.2 Staff Evaluation

The staff reviewed LRA Section 2.3.3.27 and UFSAR Sections 5.2.5.1.3, 11.5, and 12.3.4 using the evaluation methodology of SER Section 2.3. The staff conducted its review in accordance with the guidance of SRP-LR Section 2.3.

In conducting its Tier-1 review of the BOP two-tier review process, the staff evaluated the system functions described in the LRA and UFSAR to verify that the applicant had not omitted from the scope of license renewal any components with intended functions under 10 CFR 54.4(a). The staff then reviewed those components that the applicant had identified as within the scope of license renewal to verify that it had not omitted any passive and long-lived components subject to an AMR in accordance with the requirements of 10 CFR 54.21(a)(1).

The staff's review of LRA Section 2.3.3.27 identified an area in which additional information was necessary to complete the evaluation of the applicant's scoping and screening results. The applicant responded to the staff's RAI as discussed below.

In RAI 2.3.3.27-1 dated December 28, 2005, the staff noted that LRA Section 2.4.17 states that effluents through the ventilation stack are monitored to ensure that 10 CFR Part 20 limits, which apply to releases during normal operation, and 10 CFR Part 100 limits, which apply to accidental releases, are not exceeded. LRA Section 2.3.3.27 states that the stack and turbine building radioactive gaseous effluents monitors do not support a license renewal intended function and are not included within the scope of license renewal. These two statements appear to be contradictory; therefore, the staff requested that the applicant clarify this apparent contradiction and indicate whether the ventilation stack radiation monitors are within the scope of license renewal.

2-97

In its response dated January 26, 2006, the applicant stated:

> LRA Section 2.4.17 does suggest that the radiation monitors are required to monitor accident releases, but that was not the intent. While they may be used for post-accident monitoring, the stack radiation monitors are not credited for accident mitigation and are not safety-related. These radiation monitors do not have an intended function for license renewal and are therefore not in scope.

The staff review finds the applicant's response acceptable because it stated that the stack radiation monitors have no intended function for license renewal and that the LRA statement was unintentional. As such, the radiation monitors in question satisfy none of the 10 CFR 54.4(a) scoping criteria. Therefore, the staff's concern described in RAI 2.3.3.27-1 is resolved.

2.3.3.27.3 Conclusion

The staff reviewed the LRA and the RAI response to determine whether any SSCs that should be within the scope of license renewal had not been identified by the applicant. No omissions were identified. In addition, the staff determined whether any components subject to an AMR had not been identified by the applicant. No omissions were identified. The staff concludes that there is reasonable assurance that the applicant has adequately identified the radiation monitoring system components within the scope of license renewal, as required by 10 CFR 54.4(a), and those subject to an AMR, as required by 10 CFR 54.21(a)(1).

2.3.3.28 Radwaste Area Heating and Ventilation System

2.3.3.28.1 Summary of Technical Information in the Application

In LRA Section 2.3.3.28, the applicant described the radwaste area heating and ventilation system, a normally operating mechanical ventilation system to the radwaste areas of the plant including the old radwaste building, the new radwaste building, the new radwaste heat exchanger building, the offgas building, and the hot machine shop in the new maintenance building. The purpose of the system is to provide ventilation, heating, and cooling to control area temperatures, to control air movement from low contamination areas to high contamination areas, and to provide means for filtering and monitoring the exhaust air before discharging to atmosphere. It accomplishes this purpose by means of five independent HVAC systems, incorporating the necessary fans, filters, and ducting to accommodate the individual requirements of the processes within each of the five buildings. The radiological design objectives of the radwaste area heating and ventilation system are to limit the average in-plant airborne radioactivity levels below the 10 CFR Part 20 guideline limits and to reduce offsite releases of radioactivity to as low as reasonably achievable levels (10 CFR Part 50, Appendix I).

The failure of nonsafety-related SSCs in the radwaste area heating and ventilation system potentially could prevent the satisfactory accomplishment of a safety-related function.

The intended functions within the scope of license renewal include:

- provides mechanical closure

- provides pressure-retaining boundary; fission product barrier; containment isolation; or containment, holdup, and plateout (main steam system)

In LRA Table 2.3.3.28, the applicant identified the following radwaste area heating and ventilation system component types within the scope of license renewal and subject to an AMR:

- closure bolting
- damper housing
- door seal
- ductwork
- fan housing
- flexible connection

2.3.3.28.2 Staff Evaluation

The staff reviewed LRA Section 2.3.3.28 and UFSAR Sections 9.4.4 and 12.3.3 using the evaluation methodology of SER Section 2.3. The staff conducted its review in accordance with the guidance of SRP-LR Section 2.3.

The staff reviewed the subsystem functions described in the LRA and UFSAR to verify that the applicant had not omitted from the scope of license renewal any components with intended functions under 10 CFR 54.4(a). The staff then reviewed components that the applicant had identified as within the scope of license renewal to verify that it had not omitted any passive and long-lived components subject to an AMR in accordance with the requirements of 10 CFR 54.21(a)(1).

2.3.3.28.3 Conclusion

The staff reviewed the LRA to determine whether any SSCs that should be within the scope of license renewal had not been identified by the applicant. No omissions were identified. In addition, the staff determined whether any components subject to an AMR had not been identified by the applicant. No omissions were identified. The staff concludes that there is reasonable assurance that the applicant has adequately identified the radwaste area heating and ventilation system components within the scope of license renewal, as required by 10 CFR 54.4(a), and those subject to an AMR, as required by 10 CFR 54.21(a)(1).

2.3.3.29 Reactor Building Closed Cooling Water System

2.3.3.29.1 Summary of Technical Information in the Application

In LRA Section 2.3.3.29, the applicant described the reactor building closed cooling water (RBCCW) system, a closed-loop system designed to provide inhibited demineralized cooling water to reactor building and primary containment equipment subject to radioactive contamination. Included in the RBCCW system is a corrosion inhibiting chemical treatment system designed for intermittent injection of a chemical solution into the demineralized water contained within the system. The purpose of the RBCCW system is to remove heat from loads during various modes of reactor operation. The RBCCW system accomplishes this purpose by transferring heat from these loads to the service water system through the RBCCW heat exchangers. Flow and temperature control are achieved through manual/remote manipulation of RBCCW system valves. A surge tank at the high point of the system is sized to hold the expected maximum expansion of the RBCCW system. A safety injection signal (reactor vessel low-low level or drywell high pressure) trips the RBCCW pumps. Then, during operation from the

EDGs, both RBCCW pumps start automatically after a timed delay unless a LOCA signal is present. The RBCCW system acts as a buffer between radioactively contaminated systems, which it cools, and the service water system, which is the heat sink for the RBCCW system.

The RBCCW system contains safety-related components relied upon to remain functional during and following DBEs. The failure of nonsafety-related SSCs in the RBCCW system potentially could prevent the satisfactory accomplishment of a safety-related function. In addition, the RBCCW system performs functions that support fire protection and EQ.

The intended functions within the scope of license renewal include:

- provides heat transfer

- maintains mechanical and structural integrity to prevent spatial interactions that could cause failure of safety-related SSCs (includes the required structural support when the nonsafety-related leakage boundary piping is also attached to safety-related piping)

- provides mechanical closure

- provides pressure-retaining boundary; fission product barrier; or containment isolation

- provides structural support or structural integrity to preclude nonsafety-related component interactions that could prevent satisfactory accomplishment of a safety-related function

In LRA Table 2.3.3.29, the applicant identified the following RBCCW system component types within the scope of license renewal and subject to an AMR:

- closure bolting
- coolers (cleanup auxiliary pump)
- coolers (cleanup pre-coat pump)
- coolers (cleanup recirculation pumps lube oil)
- coolers (containment spray pump room)
- coolers (core spray pump room)
- coolers (drywell cooling units)
- coolers (post-accident sample)
- coolers (sample)
- coolers (shutdown cooling pumps)
- coolers (tunnel)
- filter housing
- flow element
- gauge snubber
- heat exchangers (augmented fuel pool cooling)
- heat exchangers (cleanup non-regenerative)
- heat exchangers (drywell equipment drain tank)
- heat exchangers (fuel pool cooling)
- heat exchangers (shutdown cooling)
- level glass
- piping and fittings
- pump casing (chemical feed pump)
- pump casing (RBCCW pumps)
- rupture disks

- strainer body
- tanks (chemical mixing tank)
- tanks (RBCCW surge tank)
- thermowell
- valve body

2.3.3.29.2 Staff Evaluation

The staff reviewed LRA Section 2.3.3.29 and UFSAR Sections 3.1, 9.2, 7.3, and Table 6.2-12 using the evaluation methodology of SER Section 2.3. The staff conducted its review in accordance with the guidance of SRP-LR Section 2.3.

In conducting its Tier-2 review of the BOP two-tier review process, the staff evaluated the system functions described in the LRA and UFSAR to verify that the applicant had not omitted from the scope of license renewal any components with intended functions under 10 CFR 54.4(a). The staff then reviewed those components that the applicant had identified as within the scope of license renewal to verify that it had not omitted any passive and long-lived components subject to an AMR in accordance with the requirements of 10 CFR 54.21(a)(1).

2.3.3.29.3 Conclusion

The staff reviewed the LRA to determine whether any SSCs that should be within the scope of license renewal had not been identified by the applicant. No omissions were identified. In addition, the staff determined whether any components subject to an AMR had not been identified by the applicant. No omissions were identified. The staff concludes that there is reasonable assurance that the applicant has adequately identified the RBCCW system components within the scope of license renewal, as required by 10 CFR 54.4(a), and those subject to an AMR, as required by 10 CFR 54.21(a)(1).

2.3.3.30 Reactor Building Floor and Equipment Drains

2.3.3.30.1 Summary of Technical Information in the Application

In LRA Section 2.3.3.30, the applicant described the reactor building floor and equipment drains (RFEDs). The purpose of the RFEDs is to collect floor drains and equipment drains located in the reactor building outside of the primary containment and to transfer the collected drainage to the radwaste system for processing. The RFEDs accomplish this purpose by directing floor drains first to the torus room and then to one of two sumps in the reactor building basement and directing equipment drains through a ring header to the reactor building equipment drain tank. A single pump transfers drainage from the reactor building equipment drain tank to the radwaste system collection tanks. Each level of the reactor building with the exception of the 119-foot is equipped with sufficient floor drainage capability to pass the maximum credible floor drain flow rate from actuation of the fire suppression system or a pipe break. The 119-foot level does not require a floor drain network as stairwells and equipment storage pools are sufficient to prevent flooding of this area.

The failure of nonsafety-related SSCs in the RFEDs potentially could prevent the satisfactory accomplishment of a safety-related function. The RFEDs also performs functions that support fire protection.

The intended functions within the scope of license renewal include:

- maintains mechanical and structural integrity to prevent spatial interactions that could cause failure of safety-related SSCs (includes the required structural support when the nonsafety-related leakage boundary piping is also attached to safety-related piping)
- provides mechanical closure
- provides pressure-retaining boundary

In LRA Table 2.3.3.30, the applicant identified the following RFEDs component types within the scope of license renewal and subject to an AMR:

- closure bolting
- piping and fittings
- pump casing
- tanks
- valve body

2.3.3.30.2 Staff Evaluation

The staff reviewed LRA Section 2.3.3.30 and UFSAR Sections 9.3.3 using the evaluation methodology of SER Section 2.3. The staff conducted its review in accordance with the guidance of SRP-LR Section 2.3.

In conducting its Tier-1 review of the BOP two-tier review process, the staff evaluated the system functions described in the LRA and UFSAR to verify that the applicant had not omitted from the scope of license renewal any components with intended functions under 10 CFR 54.4(a). The staff then reviewed those components that the applicant had identified as within the scope of license renewal to verify that it had not omitted any passive and long-lived components subject to an AMR in accordance with the requirements of 10 CFR 54.21(a)(1).

2.3.3.30.3 Conclusion

The staff reviewed the LRA to determine whether any SSCs that should be within the scope of license renewal had not been identified by the applicant. No omissions were identified. In addition, the staff determined whether any components subject to an AMR had not been identified by the applicant. No omissions were identified. The staff concludes that there is reasonable assurance that the applicant has adequately identified the RFEDs components within the scope of license renewal, as required by 10 CFR 54.4(a), and those subject to an AMR, as required by 10 CFR 54.21(a)(1).

2.3.3.31 Reactor Building Ventilation System

2.3.3.31.1 Summary of Technical Information in the Application

In LRA Section 2.3.3.31, the applicant described the reactor building ventilation system (RBVS), a continuously operating ventilation system with primary containment purge capability and an isolation mode. The system is designed to provide a controlled environment so that the maximum allowable ambient temperature for standard rated electrical equipment is not exceeded. It also regulates the static pressure within certain areas of the plant to minimize the spread of airborne radioactive contamination from controlled to uncontrolled areas and disposes of airborne contaminants safely. It accomplishes this regulation by maintaining a negative pressure within the reactor building as to outside atmosphere while ventilating the reactor building with fresh tempered air exhausted through the ventilation stack. The RBVS is also used during inerting and deinerting of primary containment and provides the flow paths for the SGTS and the CIS in DBEs. During normal operation, the RBVS operates with the SGTS in standby. During a DBA, the RBVS secondary containment isolation valves are closed, the RBVS fans stopped, the SGTS fans automatically started, and effluents filtered prior to elevated release through the ventilation stack.

The RBVS contains safety-related components relied upon to remain functional during and following DBEs. The failure of nonsafety-related SSCs in the RBVS potentially could prevent the satisfactory accomplishment of a safety-related function. In addition, the RBVS performs functions that support fire protection and EQ.

The intended functions within the scope of license renewal include:

- provides mechanical closure
- provides pressure-retaining boundary; fission product barrier; containment isolation; or containment, holdup, and plateout (main steam system)

In LRA Table 2.3.3.31, the applicant identified the following RBVS component types within the scope of license renewal and subject to an AMR:

- closure bolting
- closure bolting (containment isolation components)
- damper housing
- door seal
- ductwork
- piping and fittings
- piping and fittings (primary containment isolation valves)
- sensor element (temperature)
- valve body
- valve body (primary containment isolation)

2.3.3.31.2 Staff Evaluation

The staff reviewed LRA Section 2.3.3.31 and UFSAR Sections 9.4.2 and 11.3.2.5 using the evaluation methodology of SER Section 2.3. The staff conducted its review in accordance with the guidance of SRP-LR Section 2.3.

In conducting its review, the staff evaluated the system functions described in the LRA and UFSAR to verify that the applicant had not omitted from the scope of license renewal any components with intended functions under 10 CFR 54.4(a). The staff then reviewed those components that the applicant had identified as within the scope of license renewal to verify that it had not omitted any passive and long-lived components subject to an AMR in accordance with the requirements of 10 CFR 54.21(a)(1).

2.3.3.31.3 Conclusion

The staff reviewed the LRA to determine whether any SSCs that should be within the scope of license renewal had not been identified by the applicant. No omissions were identified. In addition, the staff determined whether any components subject to an AMR had not been identified by the applicant. No omissions were identified. The staff concludes that there is reasonable assurance that the applicant has adequately identified the RBVS components within the scope of license renewal, as required by 10 CFR 54.4(a), and those subject to an AMR, as required by 10 CFR 54.21(a)(1).

2.3.3.32 Reactor Water Cleanup System

2.3.3.32.1 Summary of Technical Information in the Application

In LRA Section 2.3.3.32, the applicant described the RWCU system, a filtration and demineralization system that maintains the purity of the water in the RCS. It can be operated during startup, shutdown, and refueling modes as well as during power operation.

The purposes of the RWCU system are:

- to reduce the deposition of water impurities on fuel surfaces, thus minimizing heat transfer surface fouling

- to reduce secondary sources of beta and gamma radiation by removing corrosion products, impurities, and fission products from the reactor coolant

- to reduce the concentration of chloride ions to protect steel components from chloride stress corrosion

- to maintain or lower water level in the reactor vessel during startup, shutdown, and refueling operations in order to accommodate reactor coolant swell during heatup and to accommodate water inputs from the CRD system and the head cooling system.

Portions of the RWCU System are considered RCPB. The RWCU system will automatically undergo partial or complete isolation depending upon the initiating event. Partial isolation removes the system from service without fully isolating it from the RCPB. Partial isolation will occur for RWCU system/component protection in response to RWCU system anomalies or for SLCS flow. Full isolation of the RWCU system from the RCPB occurs in response to low-low reactor water level or high drywell pressure RPS engineered safety feature system actuation parameters, or indications of an RWCU high-energy line break (HELB).

The RWCU system contains safety-related components relied upon to remain functional during and following DBEs. The failure of nonsafety-related SSCs in the RWCU system potentially could prevent the satisfactory accomplishment of a safety-related function. In addition, the RWCU system performs functions that support fire protection and EQ.

The intended functions within the scope of license renewal include:

- maintains mechanical and structural integrity to prevent spatial interactions that could cause failure of safety-related SSCs (includes the required structural support when the nonsafety-related leakage boundary piping is also attached to safety-related piping)

2-104

- provides mechanical closure
- provides pressure-retaining boundary; fission product barrier; containment isolation

In LRA Table 2.3.3.32, the applicant identified the following RWCU system component types within the scope of license renewal and subject to an AMR:

- closure bolting
- coolers (cleanup pre-coat pump)
- coolers (cleanup recirculation pumps lube oil)
- demineralizer (cleanup demineralizer)
- filter housing (cleanup filter)
- flow element
- gauge snubber
- heat exchangers (cleanup non-regenerative)
- heat exchangers (cleanup regenerative)
- piping and fittings
- pump casing (cleanup auxiliary pump)
- pump casing (cleanup filter aid pumps)
- pump casing (cleanup filter precoat pump)
- pump casing (cleanup recirc pumps)
- pump casing (cleanup sludge pump)
- restricting orifice
- sensor element
- sight glasses
- strainer body
- tanks (cleanup backwash tank)
- tanks (cleanup filter aid mix tank)
- tanks (cleanup filter and precoat tank)
- tanks (cleanup filter sludge receiver)
- tanks (cleanup recirculation pump surge tank)
- tanks (cleanup recirculation pumps lube oil)
- thermowell
- valve body

2.3.3.32.2 Staff Evaluation

The staff reviewed LRA Section 2.3.3.32 and UFSAR Sections 5.4.3, 5.4.8, and 6.2.4 using the evaluation methodology of SER Section 2.3. The staff conducted its review in accordance with the guidance of SRP-LR Section 2.3.

In conducting its Tier-2 review of the BOP two-tier review process, the staff evaluated the system functions described in the LRA and UFSAR to verify that the applicant had not omitted from the scope of license renewal any components with intended functions under 10 CFR 54.4(a). The staff then reviewed those components that the applicant had identified as within the scope of license renewal to verify that it had not omitted any passive and long-lived components subject to an AMR in accordance with the requirements of 10 CFR 54.21(a)(1).

The staff's review of LRA Section 2.3.3.32 identified an area in which additional information was necessary to complete the evaluation of the applicant's scoping and screening results. The applicant responded to the staff's RAI as discussed below.

In RAI 2.3.3.32-1 dated December 28, 2005, the staff stated that, "Note 5 on license renewal drawing LR-GE-148F444 states that the inner tube of sample cooler (at location H-8) is evaluated with the reactor water cleanup system. However, LRA Table 2.3.3.32 does not list sample cooler (tubes) as a component subject to an AMR." The staff requested that the applicant confirm that sample cooler tubes are subject to an AMR or, if not, justify their exclusion.

In its response dated January 26, 2006, the applicant stated:

> The sample cooler shown on license renewal drawing LR-GE-148F444 at drawing coordinate H-8 is a dual heat transfer coil type (tube-in-tube) with reactor building closed cooling water (RBCCW) in the annulus between the outer and inner tubes and reactor water cleanup (RWCU) water in the inner tube. Note 5 on license renewal drawing LR-GE-148F444 indicates that the inner tube of the sample cooler is evaluated with the RWCU system. The inner tube is not required for leakage boundary for license renewal as it is contained by the outer tube (which is scoped and screened with the RBCCW system). As shown on LR-GE-148F444, the inner tube is colored black indicating that the inner tube is not within the scope of license renewal (for spatial interaction) and is not subject to AMR.

The staff review finds the applicant's response acceptable because the inner tube has no potential for spatial interaction; therefore, it does not satisfy the 10 CFR 54.4(a)(2) criterion. The outer tube, which has a leakage boundary intended function, is within the scope of license renewal and subject to an AMR pursuant to 10 CFR 54.21(a)(1). Therefore, the staff's concern described in RAI 2.3.3.32-1 is resolved.

2.3.3.32.3 Conclusion

The staff reviewed the LRA and the RAI response to determine whether any SSCs that should be within the scope of license renewal had not been identified by the applicant. No omissions were identified. In addition, the staff determined whether any components subject to an AMR had not been identified by the applicant. No omissions were identified. The staff concludes that there is reasonable assurance that the applicant has adequately identified the RWCU system components within the scope of license renewal, as required by 10 CFR 54.4(a), and those subject to an AMR, as required by 10 CFR 54.21(a)(1).

2.3.3.33 Roof Drains and Overboard Discharge

2.3.3.33.1 Summary of Technical Information in the Application

In LRA Section 2.3.3.33, the applicant described the roof drains and overboard discharge system (RDODS), a passive drainage system designed to collect and discharge effluents from the plant to the discharge canal. The purpose of the RDODS is to collect and discharge effluents from plant open cooling water systems, plant building drainage systems, and yard area storm drains. The RDODS accomplishes this purpose through a 30-inch overboard discharge line that starts outside the reactor building, runs below grade, and terminates at the discharge canal. It carries service water discharge from the RBCCW heat exchangers, ESW from the containment spray system heat exchangers, turbine building sump 1 through 5 effluent, roof, floor, and equipment drainage from various plant buildings, and yard area storm water. The RDODS does not include process liquid monitoring, which is performed prior to the effluents entering the overboard discharge line. The process liquid monitoring subsystems have been designed to measure, indicate, and record the radioactivity concentration levels of major process system discharge streams continuously. These monitors assure that plant releases do no exceed the limits specified in 10 CFR Part 20 and Part 50, Appendix I.

The failure of nonsafety-related SSCs in the RDODS potentially could prevent the satisfactory accomplishment of a safety-related function. The RDODS also performs functions that support fire protection.

The intended functions within the scope of license renewal include:

- maintains mechanical and structural integrity to prevent spatial interactions that could cause failure of safety-related SSCs (includes the required structural support when the nonsafety-related leakage boundary piping is also attached to safety-related piping)
- provides mechanical closure
- provides pressure-retaining boundary

In LRA Table 2.3.3.33, the applicant identified the following RDODS component types within the scope of license renewal and subject to an AMR:

- closure bolting
- piping and fittings

2.3.3.33.2 Staff Evaluation

The staff reviewed LRA Section 2.3.3.33 and UFSAR Sections 9.3.3.2.9 using the evaluation methodology of SER Section 2.3. The staff conducted its review in accordance with the guidance of SRP-LR Section 2.3.

In conducting its Tier-1 review of the BOP two-tier review process, the staff evaluated the system functions described in the LRA and UFSAR to verify that the applicant had not omitted from the scope of license renewal any components with intended functions under 10 CFR 54.4(a). The staff then reviewed those components that the applicant had identified as within the scope of license renewal to verify that it had not omitted any passive and long-lived components subject to an AMR in accordance with the requirements of 10 CFR 54.21(a)(1).

2.3.3.33.3 Conclusion

The staff reviewed the LRA to determine whether any SSCs that should be within the scope of license renewal had not been identified by the applicant. No omissions were identified. In addition, the staff determined whether any components subject to an AMR had not been identified by the applicant. No omissions were identified. The staff concludes that there is reasonable assurance that the applicant has adequately identified the RDODS components within the scope of license renewal, as required by 10 CFR 54.4(a), and those subject to an AMR, as required by 10 CFR 54.21(a)(1).

2.3.3.34 Sanitary Waste System

2.3.3.34.1 Summary of Technical Information in the Application

In LRA Section 2.3.3.34, the applicant described the sanitary waste system, the purpose of which is to provide the path for the sanitary waste and drains to the sewage collection tank. The sanitary waste system consists of the plumbing and drainage system and the sewage lift station system. The sanitary waste system is comprised of sanitary waste piping and fixtures in the office and turbine buildings, including floor drains in the office building. Additional sanitary drains from the various plant buildings join the main sanitary drain line. Domestic waste water from all plant locations enters a concrete equalizing tank that discharges through two self-priming diaphragm pumps (transfer pumps) to the Lacey Municipal Utilities Authority sewer system and subsequently to the Ocean County Utilities Authority regional collection system via a gravity line. A radiation monitoring system continuously monitors radiation levels in the effluent of the transfer pumps. As a backup, manual samples may be taken from the sewage pit for laboratory analysis. The radiation monitor alarms below 50 percent of the 10 CFR Part 20, Appendix B, Table 1, Column 2, value for cobalt-60. Procedures require immediate notification of the control room for investigation of the alarm. If levels continue to rise, the sewage transfer pumps trip automatically below the 100 percent value identified in 10 CFR Part 20.

The failure of nonsafety-related SSCs in the sanitary waste system potentially could prevent the satisfactory accomplishment of a safety-related function.

The intended function, within the scope of license renewal, is to provide maintenance of mechanical and structural integrity to prevent spatial interactions that could cause failure of safety-related SSCs (includes the required structural support when the nonsafety-related leakage boundary piping is also attached to safety-related piping).

In LRA Table 2.3.3.34, the applicant identified the piping and fittings component type of the sanitary waste system within the scope of license renewal and subject to an AMR.

2.3.3.34.2 Staff Evaluation

The staff reviewed LRA Section 2.3.3.34 and UFSAR Sections 9.2.4.3 and 9.3.3.2.7 using the evaluation methodology of SER Section 2.3. The staff conducted its review in accordance with the guidance of SRP-LR Section 2.3.

In conducting its Tier-1 review of the BOP two-tier review process, the staff evaluated the system functions described in the LRA and UFSAR to verify that the applicant had not omitted from the scope of license renewal any components with intended functions under 10 CFR 54.4(a). The staff then reviewed those components that the applicant had identified as within the scope of license renewal to verify that it had not omitted any passive and long-lived components subject to an AMR in accordance with the requirements of 10 CFR 54.21(a)(1).

2.3.3.34.3 Conclusion

The staff reviewed the LRA to determine whether any SSCs that should be within the scope of license renewal had not been identified by the applicant. No omissions were identified. In addition, the staff determined whether any components subject to an AMR had not been identified by the applicant. No omissions were identified. The staff concludes that there is reasonable assurance that the applicant has adequately identified the sanitary waste system components within the scope of license renewal, as required by 10 CFR 54.4(a), and those subject to an AMR, as required by 10 CFR 54.21(a)(1).

2.3.3.35 Service Water System

2.3.3.35.1 Summary of Technical Information in the Application

In LRA Section 2.3.3.35, the applicant described the SWS, an open-loop cooling system designed to provide seawater to various users during normal plant operation and shutdown. The purpose of the SWS is to provide seawater cooling to the tube side of the two RBCCW heat exchangers. The SWS accomplishes this purpose by supplying seawater from the plant intake structure to the RBCCW system heat exchangers and transferring the heat energy to the environment through the RDODS. The SWS provides alternate seawater cooling to the tube side of the two TBCCW system heat exchangers normally serviced by the CWS by supplying seawater from the plant intake structure to the TBCCW system heat exchangers and transferring the heat energy to the environment through the plant discharge structure and canal. The SWS also keeps the ESW side of the containment spray heat exchangers full through a crosstie between the normally operating SWS and the standby ESW system. The SWS has several interfaces with the chlorination system, which delivers sodium hypochlorite to the SWS headers for the control of biofouling. Process liquid monitoring is for the gross radioactivity of the service water effluent from the RBCCW heat exchangers. During outages when maintenance is performed on the SWS, the ESW system can be aligned to support SWS loads through a cross-connect line between the ESW and SWS.

The failure of nonsafety-related SSCs in the SWS potentially could prevent the satisfactory accomplishment of a safety-related function. The SWS also performs functions that support fire protection.

The intended functions within the scope of license renewal include:

* provides filtration
* provides heat transfer
* maintains mechanical and structural integrity to prevent spatial interactions that could cause failure of safety-related SSCs (includes the required structural support when the nonsafety-related leakage boundary piping is also attached to safety-related piping)
* provides mechanical closure
* provides pressure-retaining boundary
* provides flow restriction

In LRA Table 2.3.3.35, the applicant identified the following SWS component types within the scope of license renewal and subject to an AMR:

* closure bolting
* eductor
* expansion joint
* flow element

- gauge snubber
- heat exchangers (RBCCW)
- heat exchangers (TBCCW)
- piping and fittings
- pump casing (rad monitor sample pump)
- pump casing (service water pumps)
- restricting orifice
- rotameter
- sample chamber
- sight glasses
- strainer
- strainer body
- tanks (service water pump oil reservoir)
- thermowell
- valve body

2.3.3.35.2 Staff Evaluation

The staff reviewed LRA Section 2.3.3.35 and UFSAR Section 9.2.1.1 using the evaluation methodology of SER Section 2.3. The staff conducted its review in accordance with the guidance of SRP-LR Section 2.3.

In conducting its Tier-2 review of the BOP two-tier review process, the staff evaluated the system functions described in the LRA and UFSAR to verify that the applicant had not omitted from the scope of license renewal any components with intended functions under 10 CFR 54.4(a). The staff then reviewed those components that the applicant had identified as within the scope of license renewal to verify that it had not omitted any passive and long-lived components subject to an AMR in accordance with the requirements of 10 CFR 54.21(a)(1).

The staff's review of LRA Section 2.3.3.35 identified an area in which additional information was necessary to complete the evaluation of the applicant's scoping and screening results. The applicant responded to the staff's RAI as discussed below.

In RAI 2.3.3.35-1 dated December 28, 2005, the staff noted that LRA Table 2.3.3.35 lists the component types "strainer" with the intended function "filter" and "strainer body" with the intended function "pressure boundary." The radiation monitor duplex strainer is indicated in parentheses for these intended functions. According to the boundaries in LRA Section 2.3.3.35 and as indicated on the license renewal drawings the following components are within the scope of license renewal and serve intended functions but are not listed in LRA Table 2.3.3.35. The staff requested that the applicant confirm that they are subject to an AMR or, if not, justify their exclusion.

(1) Strainers located at F8 and G-7 on drawing LR-BR-2005, sheet 2, that provide a pressure boundary function.

(2) The strainer S-3-035 in the seal well at B-3/4 on drawing LR-BR-2005, sheet 2, providing a filtration function. The seal well is included as part of the miscellaneous yard structures. However, there is no strainer included in this system.

In its response dated January 26, 2006, the applicant stated:

(1) The strainer symbols shown on license renewal drawing LR-BR-2005, Sheet 2 at drawing coordinates F-8 and G-7 are depicting the diaphragm seal that is integral to the pressure indicator assembly. The diaphragm seal is not specifically called out in LRA Table 2.3.3.35 since it is considered part of the "active" pressure instrument. Diaphragm seals isolate pressure instruments from the process media while allowing the instrument to sense the process pressure. A diaphragm, together with a fill fluid, transmits pressure from the process medium to the pressure element assembly of the instrument. There would be no need to filter the medium prior to the diaphragm seal. Because these diaphragm seals are part of the pressure indicator assembly, which is an "active" component, they are not subject to aging management review.

(2) Seal well strainer S-3-035 on drawing LR-BR-2005, Sheet 2 coordinate B-3/4 is incorrectly shown as in scope. This strainer was originally the supply/suction point of the service water radiation monitoring system. This strainer is no longer used and was abandoned in-place following a plant modification to the service water radiation monitoring system. This strainer does not perform an intended function for license renewal, is not in scope, and is not subject to AMR.

The staff review finds the applicant's response acceptable because the strainers in question are either parts of active components or no longer in use and satisfy none of the 10 CFR 54.4(a) criteria. Therefore, the staff's concern described in RAI 2.3.3.35-1 is resolved.

2.3.3.35.3 Conclusion

The staff reviewed the LRA and the RAI response to determine whether any SSCs that should be within the scope of license renewal had not been identified by the applicant. No omissions were identified. In addition, the staff determined whether any components subject to an AMR had not been identified by the applicant. No omissions were identified. The staff concludes that there is reasonable assurance that the applicant has adequately identified the SWS components within the scope of license renewal, as required by 10 CFR 54.4(a), and those subject to an AMR, as required by 10 CFR 54.21(a)(1).

2.3.3.36 Shutdown Cooling System

2.3.3.36.1 Summary of Technical Information in the Application

In LRA Section 2.3.3.36, the applicant described the shutdown cooling system (SCS), a high-pressure system designed to remove fission product decay heat during shutdown. The system is normally isolated and not in service during plant power operation. Immediately following shutdown of the reactor, the initial cooling and removal of decay heat is accomplished by means of the turbine bypass system, which directs steam to the main condenser. When coolant temperature has been reduced to the point where the main condenser can no longer be used as a heat sink, the SCS operates to reduce reactor coolant temperature and complete the cooling. The SCS is not an ECCS; however, the SCS may be placed in service if available during emergencies, following initial reactor cooldown and depressurization, to assist the ECCS in removing decay heat.

The SCS contains safety-related components relied upon to remain functional during and following DBEs. The failure of nonsafety-related SSCs in the SCS potentially could prevent the satisfactory accomplishment of a safety-related function. In addition, the SCS performs functions that support fire protection and EQ.

The intended functions within the scope of license renewal include:

- maintains mechanical and structural integrity to prevent spatial interactions that could cause failure of safety-related SSCs (includes the required structural support when the nonsafety-related leakage boundary piping is also attached to safety-related piping)
- provides mechanical closure
- provides pressure-retaining boundary; fission product barrier; containment isolation; or containment, holdup, and plateout (main steam system)
- provides flow restriction

In LRA Table 2.3.3.36, the applicant identified the following SCS component types within the scope of license renewal and subject to an AMR:

- closure bolting
- coolers (shutdown cooling pumps)
- flow element
- heat exchangers (shutdown cooling)
- piping and fittings
- pump casing
- restricting orifice
- thermowell
- valve body

2.3.3.36.2 Staff Evaluation

The staff reviewed LRA Section 2.3.3.36 and UFSAR Section 5.4.7 using the evaluation methodology of SER Section 2.3. The staff conducted its review in accordance with the guidance of SRP-LR Section 2.3.

In conducting its review, the staff evaluated the system functions described in the LRA and UFSAR to verify that the applicant had not omitted from the scope of license renewal any components with intended functions under 10 CFR 54.4(a). The staff then reviewed those components that the applicant had identified as within the scope of license renewal to verify that it had not omitted any passive and long-lived components subject to an AMR in accordance with the requirements of 10 CFR 54.21(a)(1).

The staff's review of LRA Section 2.3.3.36 identified an area in which additional information was necessary to complete the review of the applicant's scoping and screening results. The applicant responded to the staff's RAI as discussed below.

LRA Table 2.3.3.36 lists heat exchangers for shutdown cooling as a component type within the scope of license renewal. However, for these heat exchangers leakage/pressure boundary was identified as the sole intended function requiring aging management, not their heat transfer function. The staff believes that the heat transfer function also should be identified as an intended function of the component type and appropriate an AMP designated for reasonable assurance that this safety-related function does not degrade over the period of extended operation.

In RAI 2.3.3.36-1 dated March 10, 2006, the staff requested that the applicant clarify why the heat transfer function of the shutdown cooling heat exchangers, in addition the leakage/ pressure boundary function, had not been identified as an intended function to be preserved during the period of extended operation.

In its response dated April 7, 2006, the applicant stated that the shutdown cooling heat exchangers were identified with intended functions of heat transfer and pressure boundary in the LRA but not in Section 2.3.3.36. The subject components were listed in LRA Table 2.3.3.36 as requiring an AMR without heat removal as an intended function because heat removal is not credited as a 10 CFR 54.4(a)(1) function. However, the system is relied upon for a function for compliance with 10 CFR 54.4(a)(3) for fire protection. Consequently, the shutdown cooling heat exchangers are listed in LRA Table 2.3.3.29 for RBCCW system components subject to an AMR and with both intended functions of heat transfer and pressure boundary.

The staff finds the response acceptable as a clarification. The staff's concern described in RAI 2.3.3.36-1 is resolved.

2.3.3.36.3 Conclusion

The staff reviewed the LRA and the RAI response to determine whether any SSCs that should be within the scope of license renewal had not been identified by the applicant. No omissions were identified. In addition, the staff determined whether any components subject to an AMR had not been identified by the applicant. No omissions were identified. The staff concludes that there is reasonable assurance that the applicant has adequately identified the SCS components within the scope of license renewal, as required by 10 CFR 54.4(a), and those subject to an AMR, as required by 10 CFR 54.21(a)(1).

2.3.3.37 Spent Fuel Pool Cooling System

2.3.3.37.1 Summary of Technical Information in the Application

In LRA Section 2.3.3.37, the applicant described the spent fuel pool cooling system (SFPCS), which consists of two systems located in the reactor building that operate independently from each other except for a common suction flow path and a common discharge flow path. The first system is the SFPCS designed to remove heat from the spent fuel pool and maintain fuel storage pool water clarity. The other system is the augmented SFPCS added after plant construction due to higher than anticipated spent fuel storage requirements. This system operates during refueling due to the higher heat loads. The SFPCS is designed for both normal and accident conditions of loss of offsite power coincident with a single active component failure. The augmented SFPCS is designed to provide a seismically qualified cooling loop capable of providing cooling during such conditions. The system is designed to prevent reduction in fuel storage coolant inventory during accident conditions. In addition, the system is designed with sufficient monitoring systems to detect conditions that could cause loss of decay heat removal and to initiate appropriate safety actions. Telltale drains with annunciated flow-indicating switches detect leakage through the bellows seal at the reactor vessel to drywell joint and leakage into the space between the refueling gates. There is a curb around the cavities to direct any overflow to drains.

The SFPCS has safety-related components relied upon to remain functional during and following DBEs. The failure of nonsafety-related SSCs in the SFPCS potentially could prevent the satisfactory accomplishment of a safety-related function.

The intended functions within the scope of license renewal include:

- provides spray shield or curbs for directing flow
- maintains mechanical and structural integrity to prevent spatial interactions that could cause failure of safety-related SSCs (includes the required structural support when the nonsafety-related leakage boundary piping is also attached to safety-related piping)
- provides mechanical closure
- provides pressure-retaining boundary; fission product barrier; containment isolation

In LRA Table 2.3.3.37, the applicant identified the following SFPCS component types within the scope of license renewal and subject to an AMR:

- closure bolting
- diffuser
- flow element
- piping and fittings
- pump casing (fuel pool cooling pumps and augmented fuel pool cooling pumps)
- thermowells
- valve body

2.3.3.37.2 Staff Evaluation

The staff reviewed LRA Section 2.3.3.37 and UFSAR Sections 1.2, 3.1, 3.2, 7.5, 9.1, and 11.1 using the evaluation methodology of SER Section 2.3. The staff conducted its review in accordance with the guidance of SRP-LR Section 2.3.

In conducting its Tier-2 review of the BOP two-tier review process, the staff evaluated the system functions described in the LRA and UFSAR to verify that the applicant had not omitted from the scope of license renewal any components with intended functions under 10 CFR 54.4(a). The staff then reviewed those components that the applicant had identified as within the scope of license renewal to verify that it had not omitted any passive and long-lived components subject to an AMR in accordance with the requirements of 10 CFR 54.21(a)(1).

The staff's review of LRA Section 2.3.3.37 identified an area in which additional information was necessary to complete the evaluation of the applicant's scoping and screening results. The applicant responded to the staff's RAI as discussed below.

In RAI 2.3.3.37-1 dated December 28, 2005, the staff noted that LRA Section 2.3.3.37 states that the piping that discharges into the reactor cavity, equipment storage cavity, and spent fuel pool is included in the scoping boundary for the SPFCS. However, drawing LR-GE-237E756 (location E-9) does not highlight the piping and diffusers that discharge into the reactor cavity as within the scoping boundary. Therefore, the staff requested that the applicant clarify this discrepancy.

In its response dated January 26, 2006, the applicant stated:

> The piping and return diffusers located within the reactor cavity are correctly shown on license renewal Drawing LR-GE-237E756 as not in scope (black). The piping up to the reactor cavity is in scope, but the piping within the reactor cavity does not perform or support a system intended function. The intent of the discussion in LRA Section 2.3.3.37 was not to exactly define the components or portion of piping that was in scope, but rather to describe this section in general terms. The exact boundary of in scope/not in scope piping is defined by the license renewal drawing.

The staff review finds the applicant's response acceptable because the piping and diffusers that discharge into the reactor cavity support or perform no system intended function and satisfy no 10 CFR 54.4(a) criteria. Therefore, the staff's concern described in RAI 2.3.3.37-1 is resolved.

2.3.3.37.3 Conclusion

The staff reviewed the LRA and the RAI response to determine whether any SSCs that should be within the scope of license renewal had not been identified by the applicant. No omissions were identified. In addition, the staff determined whether any components subject to an AMR had not been identified by the applicant. No omissions were identified. The staff concludes that there is reasonable assurance that the applicant has adequately identified the SFPCS components within the scope of license renewal, as required by 10 CFR 54.4(a), and those subject to an AMR, as required by 10 CFR 54.21(a)(1).

2.3.3.38 Standby Liquid Control System (Liquid Poison System)

2.3.3.38.1 Summary of Technical Information in the Application

In LRA Section 2.3.3.38, the applicant described the standby liquid control system (SLCS) or the liquid poison system, a standby and redundant sodium pentaborate injection system designed to bring the reactor to a shutdown condition at any time in core life independent of control rod capabilities. The SLCS operates independently from the CRD system. The most severe requirement for which the system is designed is shutdown from a full power operating condition assuming complete failure of the CRD system to respond to a scram signal. The SLCS provides sufficient capacity for controlling the reactivity difference between the steady state rated operating condition of the reactor and the cold shutdown condition, including shutdown margin, thereby ensuring complete shutdown capability from the most reactive condition at any time in core life. The SLCS accomplishes this purpose by injecting sodium pentaborate solution into the reactor vessel to absorb thermal neutrons. The SLCS is not provided as a backup for reactor trip functions, since most transient conditions requiring reactor trip occur too rapidly to be controlled by the SLCS. The SLCS is manually initiated from the main control room through the use of a keylock switch to start the selected pump and actuate its explosive actuated valve. This manual initiation ensures that switching on the system is a deliberate act. Following system initiation, the explosive valve of the selected pump is actuated to provide a flow path to the reactor vessel.

The SLCS contains safety-related components relied upon to remain functional during and following DBEs. The failure of nonsafety-related SSCs in the SLCS potentially could prevent the satisfactory accomplishment of a safety-related function. In addition, the SLCS performs functions that support ATWS.

The intended functions within the scope of license renewal include:

- maintains mechanical and structural integrity to prevent spatial interactions that could cause failure of safety-related SSCs (includes the required structural support when the nonsafety-related leakage boundary piping is also attached to safety-related piping)
- provides mechanical closure
- provides pressure-retaining boundary; fission product barrier; containment isolation; or containment, holdup, and plateout (main steam system)
- provides structural support or structural integrity to preclude nonsafety-related component interactions that could prevent satisfactory accomplishment of a safety-related function

In LRA Table 2.3.3.38, the applicant identified the following SLCS component types within the scope of license renewal and subject to an AMR:

- accumulator
- closure bolting
- flow element
- piping and fittings
- pump casing
- tanks (liquid poison tank)
- tanks (liquid poison test tank)
- thermowell
- valve body

2.3.3.38.2 Staff Evaluation

The staff reviewed LRA Section 2.3.3.38 and UFSAR Sections 3.1, 4.6.4.1, 7.4.1, 9.3.5, and 15.8 using the evaluation methodology of SER Section 2.3. The staff conducted its review in accordance with the guidance of SRP-LR Section 2.3.

In conducting its review, the staff evaluated the system functions described in the LRA and UFSAR to verify that the applicant had not omitted from the scope of license renewal any components with intended functions under 10 CFR 54.4(a). The staff then reviewed those components that the applicant had identified as within the scope of license renewal to verify that it had not omitted any passive and long-lived components subject to an AMR in accordance with the requirements of 10 CFR 54.21(a)(1).

2.3.3.38.3 Conclusion

The staff reviewed the LRA to determine whether any SSCs that should be within the scope of license renewal had not been identified by the applicant. No omissions were identified. In addition, the staff determined whether any components subject to an AMR had not been identified by the applicant. No omissions were identified. The staff concludes that there is reasonable assurance that the applicant has adequately identified the SLCS components within the scope of license renewal, as required by 10 CFR 54.4(a), and those subject to an AMR, as required by 10 CFR 54.21(a)(1).

2.3.3.39 Traveling In-Core Probe System

2.3.3.39.1 Summary of Technical Information in the Application

In LRA Section 2.3.3.39, the applicant described the TIP system, an electrical instrumentation system designed to provide neutron flux data for calibration of the local power range monitor (LPRM) detectors and determination of axial neutron flux levels for core power distribution measurements. The purpose of the TIP system is to measure core neutron flux at various positions throughout the core. The TIP system accomplishes its purpose by utilizing a set of fission chamber detector instruments identical to those used by the LPRM system and a positioning system capable of moving the fission chamber detectors to various locations in the core corresponding to the locations of the LPRM detectors. The moveable TIP detectors, as with the fixed LPRM detectors, generate signals processed to indicate neutron flux levels in the vicinity of each detector. As the TIP detectors may be fully withdrawn from the core and outside of primary containment, the TIP system contains mechanical components designed to assure primary containment integrity. The TIP system does not generate any rod block or scram signals for protection of the reactor; however, the portion responsible for providing primary containment integrity is within the scope for license renewal.

The TIP system contains safety-related components relied upon to remain functional during and following DBEs.

The intended functions within the scope of license renewal include:

*   provides mechanical closure
*   provides pressure-retaining boundary; fission product barrier; containment isolation; or containment, holdup, and plateout (main steam system)

In LRA Table 2.3.3.39, the applicant identified the following TIP system component types within the scope of license renewal and subject to an AMR:

*   closure bolting
*   piping and fittings
*   valve body

2.3.3.39.2 Staff Evaluation

The staff reviewed LRA Section 2.3.3.39 and UFSAR Section 7.5.1.8.8 using the evaluation methodology of SER Section 2.3. The staff conducted its review in accordance with the guidance of SRP-LR Section 2.3.

In conducting its review, the staff evaluated the system functions described in the LRA and UFSAR to verify that the applicant had not omitted from the scope of license renewal any components with intended functions under 10 CFR 54.4(a). The staff then reviewed those components that the applicant had identified as within the scope of license renewal to verify that it had not omitted any passive and long-lived components subject to an AMR in accordance with the requirements of 10 CFR 54.21(a)(1).

2.3.3.39.3 Conclusion

The staff reviewed the LRA to determine whether any SSCs that should be within the scope of license renewal had not been identified by the applicant. No omissions were identified. In addition, the staff determined whether any components subject to an AMR had not been identified by the applicant. No omissions were identified. The staff concludes that there is reasonable assurance that the applicant has adequately identified the TIP system components within the scope of license renewal, as required by 10 CFR 54.4(a), and those subject to an AMR, as required by 10 CFR 54.21(a)(1).

2.3.3.40 Turbine Building Closed Cooling Water System

2.3.3.40.1 Summary of Technical Information in the Application

In LRA Section 2.3.3.40, the applicant described the TBCCW system, a closed-loop system designed to provide inhibited demineralized cooling water to the reactor recirculation pump MG sets and turbine building equipment not subject to radioactive contamination. Included in the TBCCW system is a corrosion-inhibiting chemical treatment system designed for intermittent injection of a chemical solution into the demineralized water within the system. The purpose of the TBCCW system is to remove heat from various loads during all modes of reactor operation. The TBCCW system accomplishes this purpose by transferring heat from these loads to either the CWS (normal cooling water supply to TBCCW heat exchangers) or the SWS (alternate cooling supply to TBCCW heat exchangers) through the TBCCW heat exchangers. Except for TBCCW flow to the hydrogen coolers, all system valving is manual. TBCCW flow to the hydrogen coolers is through an air-operated valve that can be operated in a temperature-regulated automatic or manual mode.

The failure of nonsafety-related SSCs in the TBCCW system potentially could prevent the satisfactory accomplishment of a safety-related function.

The intended functions within the scope of license renewal include:

- maintains mechanical and structural integrity to prevent spatial interactions that could cause failure of safety-related SSCs (includes the required structural support when the nonsafety-related leakage boundary piping is also attached to safety-related piping)

- provides mechanical closure

In LRA Table 2.3.3.40, the applicant identified the following TBCCW system component types within the scope of license renewal and subject to an AMR:

- closure bolting
- coolers (condensate pump motor)
- coolers (condenser vacuum pump)
- coolers (control room AC)
- coolers (feedwater and main steam sample)

- coolers (feedwater pump lube oil)
- coolers (final feedwater facility)
- coolers (hydrogen)
- coolers (reactor recirculation pump M-G sets)
- coolers (service air compressor aftercooler)
- coolers (service air compressor cylinders)
- coolers (service air compressor intercooler)
- coolers (stator winding liquid)
- coolers (thermal control unit)
- coolers (turbine lube oil)
- filter housing
- flexible connection
- flow element
- flow glass
- gauge snubber
- heat exchangers (generator bus)
- heat exchangers (TBCCW)
- level glass
- piping and fittings
- pump casing (TBCCW pumps, chemical feed pump)
- strainer body
- tanks (surge, chemical mixing, closed cooling water)
- thermowell
- valve body

2.3.3.40.2 Staff Evaluation

The staff reviewed LRA Section 2.3.3.40 and UFSAR Sections 9.2, 10.4, 5.4, and 9.1 using the evaluation methodology of SER Section 2.3. The staff conducted its review in accordance with the guidance of SRP-LR Section 2.3.

In conducting its Tier-2 review of the BOP two-tier review process, the staff evaluated the system functions described in the LRA and UFSAR to verify that the applicant had not omitted from the scope of license renewal any components with intended functions under 10 CFR 54.4(a). The staff then reviewed those components that the applicant had identified as within the scope of license renewal to verify that it had not omitted any passive and long-lived components subject to an AMR in accordance with the requirements of 10 CFR 54.21(a)(1).

2.3.3.40.3 Conclusion

The staff reviewed the LRA to determine whether any SSCs that should be within the scope of license renewal had not been identified by the applicant. No omissions were identified. In addition, the staff determined whether any components subject to an AMR had not been identified by the applicant. No omissions were identified. The staff concludes that there is reasonable assurance that the applicant has adequately identified the TBCCW system components within the scope of license renewal, as required by 10 CFR 54.4(a), and those subject to an AMR, as required by 10 CFR 54.21(a)(1).

2.3.3.41 Water Treatment & Distribution System

2.3.3.41.1 Summary of Technical Information in the Application

In LRA Section 2.3.3.41, the applicant described the water treatment and distribution system, the purpose of which is to be the source of all potable water, demineralized water, and condensate for the station. It accomplishes this purpose by drawing fresh water from a deep well for processing in the pretreatment system. After treatment, part of the water goes to the domestic water system and the rest is further treated in the makeup demineralizer (MUD) system. The water treatment and distribution system consists of the following subsystems: pretreatment subsystem, domestic water and domestic water distribution subsystem, MUD subsystem, and demineralized water transfer subsystem. The pretreatment subsystem is trailer-mounted and designed to filter the raw water drawn from the well pit by the deep well pumps. The domestic water subsystem is designed to provide a supply of fresh water for use by all site facilities including laundry, drinking fountains, kitchens, bathrooms, eye wash stations, decontamination showers, HVAC (air washers and SEB computer room), select sump pump bearing coolers, and the MUD subsystem. The domestic water subsystem consists of two subsystems, the original domestic water subsystem and the north yard domestic water subsystem. The domestic water distribution subsystem is designed to distribute potable water throughout the facility. A chemical feed subsystem treats the original domestic water prior to use. The MUD subsystem is designed to take pretreated water from the domestic water system and process it to meet the high purity standards of water for makeup purposes. The original MUD subsystem was replaced by a mobile demineralizer unit for purifying filtered well water before transfer to the demineralized water storage tank (DWST). The demineralized water transfer subsystem is designed to store demineralized water in the DWST and to supply an adequate amount for various plant uses. The demineralized water transfer subsystem is normally kept in operation at all times. During a loss of offsite power, either transfer pump may be started manually and operated from the EDGs if there is a demand on the system.

The water treatment and distribution system contains safety-related components relied upon to remain functional during and following DBEs. The failure of nonsafety-related SSCs in the water treatment and distribution system potentially could prevent the satisfactory accomplishment of a safety-related function.

The intended functions within the scope of license renewal include:

- maintains mechanical and structural integrity to prevent spatial interactions that could cause failure of safety-related SSCs (includes the required structural support when the nonsafety-related leakage boundary piping is also attached to safety-related piping)
- provides mechanical closure
- provides pressure-retaining boundary

In LRA Table 2.3.3.41, the applicant identified the following water treatment and distribution system component types within the scope of license renewal and subject to an AMR:

- closure bolting
- filter housing (including purifier M-12-1)
- flexible hose
- flow element

- flow meter
- piping and fittings
- restricting orifice
- tanks (including hot water heater H-12-1)
- valve body

2.3.3.41.2 Staff Evaluation

The staff reviewed LRA Section 2.3.3.41 and UFSAR Sections 9.2.3 and 6.4.2.1 using the evaluation methodology of SER Section 2.3. The staff conducted its review in accordance with the guidance of SRP-LR Section 2.3.

In conducting its Tier-2 review of the BOP two-tier review process, the staff evaluated the system functions described in the LRA and UFSAR to verify that the applicant had not omitted from the scope of license renewal any components with intended functions under 10 CFR 54.4(a). The staff then reviewed those components that the applicant had identified as within the scope of license renewal to verify that it had not omitted any passive and long-lived components subject to an AMR in accordance with the requirements of 10 CFR 54.21(a)(1).

2.3.3.41.3 Conclusion

The staff reviewed the LRA to determine whether any SSCs that should be within the scope of license renewal had not been identified by the applicant. No omissions were identified. In addition, the staff determined whether any components subject to an AMR had not been identified by the applicant. No omissions were identified. The staff concludes that there is reasonable assurance that the applicant has adequately identified the water treatment and distribution system components within the scope of license renewal, as required by 10 CFR 54.4(a), and those subject to an AMR, as required by 10 CFR 54.21(a)(1).

2.3.4 Steam and Power Conversion Systems

In LRA Section 2.3.4, the applicant identified the SCs of the steam and power conversion systems subject to an AMR for license renewal.

The applicant described the supporting SCs of the steam and power conversion systems in the following sections of the LRA:

- 2.3.4.1   condensate system
- 2.3.4.2   condensate transfer system
- 2.3.4.3   feedwater system
- 2.3.4.4   main condenser
- 2.3.4.5   main generator and auxiliary system
- 2.3.4.6   main steam system
- 2.3.4.7   main turbine and auxiliary system

The staff findings on LRA Sections 2.3.4.1 – 2.3.4.7 are presented in SER Sections 2.3.4.1 – 2.3.4.7, respectively.

2.3.4.1 Condensate System

2.3.4.1.1 Summary of Technical Information in the Application

In LRA Section 2.3.4.1, the applicant described the condensate system (CNDS) designed to transfer sub-cooled condensate from the main condenser hotwell to the feedwater system. It to transfers condensate water from the main condenser through the condensate demineralizer and supplies the reactor feed pump at a suitable pressure and required purity level. The CNDS includes the condensate system and the condensate demineralizer system. During normal plant operations, the purpose of the CNDS is to purify condensate by removing corrosion products, dissolved solids, chemicals, and other impurities that may enter the reactor coolant cycle. The CNDS accomplishes this purpose by processing the condensate through demineralizers. In the likely event that station auxiliary power is available, the condensate and feedwater systems provide additional emergency core cooling capability.

The failure of nonsafety-related SSCs in the CNDS potentially could prevent the satisfactory accomplishment of a safety-related function.

The intended functions within the scope of license renewal include:

- maintains mechanical and structural integrity to prevent spatial interactions that could cause failure of safety-related SSCs (includes the required structural support when the nonsafety-related leakage boundary piping is also attached to safety-related piping)
- provides mechanical closure

In LRA Table 2.3.4.1, the applicant identified the following CNDS component types within the scope of license renewal and subject to an AMR:

- closure bolting
- expansion joint
- filter housing
- flow element
- heat exchangers
- piping and fittings
- pump casing
- restricting orifice
- sensor element
- sight glasses
- strainer body
- tanks
- thermowell
- valve body

2.3.4.1.2 Staff Evaluation

The staff reviewed LRA Section 2.3.4.1 and UFSAR Sections 10.1, 10.4.6, and 10.4.7 using the evaluation methodology of SER Section 2.3. The staff conducted its review in accordance with the guidance of SRP-LR Section 2.3.

In conducting its Tier-1 review of the BOP two-tier review process, the staff evaluated the system functions described in the LRA and UFSAR to verify that the applicant had not omitted from the scope of license renewal any components with intended functions under 10 CFR 54.4(a). The staff then reviewed those components that the applicant had identified as within the scope of license renewal to verify that it had not omitted any passive and long-lived components subject to an AMR in accordance with the requirements of 10 CFR 54.21(a)(1).

2.3.4.1.3 Conclusion

The staff reviewed the LRA to determine whether any SSCs that should be within the scope of license renewal had not been identified by the applicant. No omissions were identified. In addition, the staff determined whether any components subject to an AMR had not been identified by the applicant. No omissions were identified. The staff concludes that there is reasonable assurance that the applicant has adequately identified the CNDS components within the scope of license renewal, as required by 10 CFR 54.4(a), and those subject to an AMR, as required by 10 CFR 54.21(a)(1).

2.3.4.2 Condensate Transfer System

2.3.4.2.1 Summary of Technical Information in the Application

In LRA Section 2.3.4.2, the applicant described the condensate transfer system, a condensate storage, makeup, and supply system designed to distribute water to the control rod drive, core spray, condensate, isolation condenser, reactor water clean up, spent fuel pool cooling, radwaste and the heater, drains, and vent and pressure systems. The purpose of the condensate transfer system is to provide bulk storage of condensate, surge volume capability for the condensate system, condensate supply for the condensate demineralizer resin transfer, flushing, resin regeneration, and makeup to the isolation condensers and spent fuel pool. Condensate is also supplied by the condensate transfer system for pump bearing cooling and makeup supply for various plant systems. It accomplishes these purposes by continuously delivering condensate from the condensate transfer pumps to individual plant systems. It also provides a flow path between plant water supplies and various pumps and equipment when the appropriate manual or remote manual line-ups are made. The system is normally filled by the demineralized water transfer system and has an emergency fill from the fire protection system. The system operates continuously during plant power operation and is credited to support the isolation condensers for plant shutdown.

The condensate transfer system contains safety-related components relied upon to remain functional during and following DBEs. The failure of nonsafety-related SSCs in the condensate transfer system potentially could prevent the satisfactory accomplishment of a safety-related function. In addition, the condensate transfer system performs functions that support fire protection and SBO.

The intended functions within the scope of license renewal include:

- maintains mechanical and structural integrity to prevent spatial interactions that could cause failure of safety-related SSCs (includes the required structural support when the nonsafety-related leakage boundary piping is also attached to safety-related piping)

- provides mechanical closure

- provides pressure-retaining boundary
- provides flow restriction

In LRA Table 2.3.4.2, the applicant identified the following condensate transfer system component types within the scope of license renewal and subject to an AMR:

- closure bolting
- expansion joint
- flow element
- gauge snubber
- piping and fittings
- pump casing
- restricting orifice
- tanks
- valve body

2.3.4.2.2 Staff Evaluation

The staff reviewed LRA Section 2.3.4.2 and UFSAR Sections 10.4.7, 7.4, 6.3, 15.2.6, and 9.1 using the evaluation methodology of SER Section 2.3. The staff conducted its review in accordance with the guidance of SRP-LR Section 2.3.

In conducting its Tier-2 review of the BOP two-tier review process, the staff evaluated the system functions described in the LRA and UFSAR to verify that the applicant had not omitted from the scope of license renewal any components with intended functions under 10 CFR 54.4(a). The staff then reviewed those components that the applicant had identified as within the scope of license renewal to verify that it had not omitted any passive and long-lived components subject to an AMR in accordance with the requirements of 10 CFR 54.21(a)(1).

2.3.4.2.3 Conclusion

The staff reviewed the LRA to determine whether any SSCs that should be within the scope of license renewal had not been identified by the applicant. No omissions were identified. In addition, the staff determined whether any components subject to an AMR had not been identified by the applicant. No omissions were identified. The staff concludes that there is reasonable assurance that the applicant has adequately identified the condensate transfer system components within the scope of license renewal, as required by 10 CFR 54.4(a), and those subject to an AMR, as required by 10 CFR 54.21(a)(1).

2.3.4.3 Feedwater System

2.3.4.3.1 Summary of Technical Information in the Application

In LRA Section 2.3.4.3, the applicant described the feedwater system, a reactor water level control system that provides reheated condensate water to the RPV during normal operation at a flow rate equivalent to what is generated into steam by boil-off and removed by the main steam system. Essential for power operations, the feedwater system provides cooling water to the core during a LOCA but is not credited in accident analyses, not considered part of the ECCS, nor credited to support safe shutdown. The feedwater system includes the feedwater control system, the reactor feed pump lube oil system, and the zinc injection system. The feedwater control system is a digital control function of the feedwater system. Reactor water level is controlled by the positions of the low flow or main feedwater regulating valves controlling feedwater flow rate to the reactor vessel. The zinc injection system injects depleted zinc oxide into the RCS to reduce deposits and shutdown dose rates in RCS piping and components.

The feedwater system contains safety-related components relied upon to remain functional during and following DBEs. The failure of nonsafety-related SSCs in the feedwater system potentially could prevent the satisfactory accomplishment of a safety-related function. In addition, the feedwater system performs functions that support fire protection.

The intended functions within the scope of license renewal include:

- maintains mechanical and structural integrity to prevent spatial interactions that could cause failure of safety-related SSCs (includes the required structural support when the nonsafety-related leakage boundary piping is also attached to safety-related piping)
- provides mechanical closure
- provides pressure-retaining boundary or containment isolation

In LRA Table 2.3.4.3, the applicant identified the following feedwater system component types within the scope of license renewal and subject to an AMR:

- closure bolting
- dissolution column
- expansion joint
- filter housing
- flow element
- heat exchangers
- piping and fittings
- pump casing
- strainer body
- tanks
- thermowell
- valve body

2.3.4.3.2 Staff Evaluation

The staff reviewed LRA Section 2.3.4.3 and UFSAR Sections 7.6.1.1, 7.7.1.4, 10.1, 10.4.7, and 15.1 using the evaluation methodology of SER Section 2.3. The staff conducted its review in accordance with the guidance of SRP-LR Section 2.3.

In conducting its Tier-2 review of the BOP two-tier review process, the staff evaluated the system functions described in the LRA and UFSAR to verify that the applicant had not omitted from the scope of license renewal any components with intended functions under 10 CFR 54.4(a). The staff then reviewed those components that the applicant had identified as within the scope of license renewal to verify that it had not omitted any passive and long-lived components subject to an AMR in accordance with the requirements of 10 CFR 54.21(a)(1).

The staff's review of LRA Section 2.3.4.3 identified an area in which additional information was necessary to complete the evaluation of the applicant's scoping and screening results. The applicant responded to the staff's RAI as discussed below.

In RAI 2.3.4.3-1 dated December 28, 2005, the staff noted that, although LRA Section 2.3.4.3 includes the feedwater system within the scope of license renewal for a fire protection intended function, in its license renewal drawing of components with intended functions it is not obvious which feedwater system components actually are credited with a fire protection intended function in accordance with 10 CFR 54.4(a)(3). Therefore, the staff requested that the applicant identify those portions of the feedwater system with fire protection functions required for 10 CFR 54.4(a)(3).

In its response dated January 26, 2006, the applicant stated:

> LRA Section 2.3.4.3 for the feedwater system does not specifically identify the portion of the system relied upon for fire protection. The feedwater control system, which is included in the feedwater system license renewal system, is the portion relied upon for fire protection. The feedwater control system is not shown on license renewal drawing LR-BR-2003 for feedwater.
>
> The feedwater control system provides a digital control function for the feedwater system and consists of two computers with dual links to the digital controllers. The computers contain the feedwater logic software. The Appendix R safe shutdown analysis requires demonstration of adequate plant process monitoring capability to achieve and maintain safe shutdown during and following postulated fire events. The Oyster Creek safe shutdown analysis credits reactor level monitoring instrumentation, including associated control and indication circuits that are part of the feedwater control system.

The staff review finds the applicant's response acceptable because it identified the portions of the feedwater system relied upon for fire protection in accordance with 10 CFR 54.4(a)(3). Therefore, the staff's concern described in RAI 2.3.4.3-1 is resolved.

2.3.4.3.3 Conclusion

The staff reviewed the LRA and the RAI response to determine whether any SSCs that should be within the scope of license renewal had not been identified by the applicant. No omissions were identified. In addition, the staff determined whether any components subject to an AMR had not been identified by the applicant. No omissions were identified. The staff concludes that there is reasonable assurance that the applicant has adequately identified the Feedwater system components within the scope of license renewal, as required by 10 CFR 54.4(a), and those subject to an AMR, as required by 10 CFR 54.21(a)(1).

2.3.4.4 Main Condenser

2.3.4.4.1 Summary of Technical Information in the Application

In LRA Section 2.3.4.4, the applicant described the main condenser, a heat sink for the turbine exhaust steam, turbine bypass steam, and other flows. It also deaerates and stores the condensate for reuse after a period of radioactive decay. Additionally, the main condenser provides for post-accident containment, holdup, and plateout of MSIV bypass leakage.

The main condenser is designed to:

(1) accept a portion of turbine bypass steam flow without exceeding the turbine exhaust pressure and temperature limitations

(2) receive, in addition to the main turbine exhaust, vents and drains from the regenerative feedwater heating system and from various other components and systems of the heat cycle

(3) provide time for radioactive isotope decay by retaining sufficient water in the hotwell without makeup and with turbine throttle valves wide open

The purpose of the system is to condense low-pressure turbine exhaust from each of the low-pressure turbines and allow for the decay of short-lived isotopes. The main condenser accomplishes this purpose by transferring heat to the circulating water system and by ensuring sufficient retention time in the hotwell to allow for the decay of short-lived isotopes.

The failure of nonsafety-related SSCs in the main condenser potentially could prevent the satisfactory accomplishment of a safety-related function.

The intended function, within the scope of license renewal, is to provide post-accident containment, plateout of iodine, and hold-up of iodine and noncondensible gases before release.

In LRA Table 2.3.4.4, the applicant identified the following main condenser component types within the scope of license renewal and subject to an AMR:

- main condenser shell
- main condenser tubes
- main condenser tubesheet

2.3.4.4.2 Staff Evaluation

The staff reviewed LRA Section 2.3.4.4 and UFSAR Section 10.4.1 using the evaluation methodology of SER Section 2.3. The staff conducted its review in accordance with the guidance of SRP-LR Section 2.3.

In conducting its Tier-1 review of the BOP two-tier review process, the staff evaluated the system functions described in the LRA and UFSAR to verify that the applicant had not omitted from the scope of license renewal any components with intended functions under 10 CFR 54.4(a). The staff then reviewed those components that the applicant had identified as within the scope of license renewal to verify that it had not omitted any passive and long-lived components subject to an AMR in accordance with the requirements of 10 CFR 54.21(a)(1).

2.3.4.4.3 Conclusion

The staff reviewed the LRA to determine whether any SSCs that should be within the scope of license renewal had not been identified by the applicant. No omissions were identified. In addition, the staff determined whether any components subject to an AMR had not been identified by the applicant. No omissions were identified. The staff concludes that there is reasonable assurance that the applicant has adequately identified the main condenser components within the scope of license renewal, as required by 10 CFR 54.4(a), and those subject to an AMR, as required by 10 CFR 54.21(a)(1).

2.3.4.5 Main Generator and Auxiliary System

2.3.4.5.1 Summary of Technical Information in the Application

In LRA Section 2.3.4.5, the applicant described the main generator and auxiliary system (MGAS), a normally operating system designed to convert the mechanical energy of the turbine into electrical energy fed to the main transmission lines and also used to satisfy in-house loads. The MGAS is comprised of the following subsystems: main generator, main generator exciter, stator cooling, hydrogen cooling, hydrogen seal oil, and the generator isolated phase bus. The main generator consists of a casing, a rotor, and a stator. The casing forms a gas-tight boundary. The rotor consists of the rotor body with two shaft extensions. Hydrogen flows into the rotor near each retaining ring to cool the copper windings. Two axial blower-type fans, one at each end of the rotor, circulate cooling hydrogen gas around the generator and through the coolers. The stator contains the main generator armature windings and consists of the stator core and stator windings. The stator windings are directly water-cooled by stator cooling water which removes heat produced in the stator bars of the main generator. The main exciter supplies the main generator field with excitation voltage through a slip ring/brush rigging arrangement and the main exciter output circuit breaker. The hydrogen seal oil subsystem maintains the hydrogen inside the generator casing. The isolated phase bus connects the main generator to the main transformers, auxiliary transformer, and generator neutral connection.

The failure of nonsafety-related SSCs in the MGAS potentially could prevent the satisfactory accomplishment of a safety-related function.

The intended functions within the scope of license renewal include:

- maintains mechanical and structural integrity to prevent spatial interactions that could cause failure of safety-related SSCs (includes the required structural support when the nonsafety-related leakage boundary piping is also attached to safety-related piping)
- provides mechanical closure

In LRA Table 2.3.4.5, the applicant identified the following MGAS component types within the scope of license renewal and subject to an AMR:

- closure bolting
- filter housing
- flow element
- gauge snubber
- heat exchangers
- piping and fittings
- pump casing
- restricting orifice
- sensor element
- sight glasses
- strainer body
- tanks
- valve body

2.3.4.5.2 Staff Evaluation

The staff reviewed LRA Section 2.3.4.5 and UFSAR Sections 8.1.2, 8.2.1, 8.3.1.1, 9.2.1, and 10.2.2 using the evaluation methodology of SER Section 2.3. The staff conducted its review in accordance with the guidance of SRP-LR Section 2.3.

In conducting its Tier-1 review of the BOP two-tier review process, the staff evaluated the system functions described in the LRA and UFSAR to verify that the applicant had not omitted from the scope of license renewal any components with intended functions under 10 CFR 54.4(a). The staff then reviewed those components that the applicant had identified as within the scope of license renewal to verify that it had not omitted any passive and long-lived components subject to an AMR in accordance with the requirements of 10 CFR 54.21(a)(1).

2.3.4.5.3 Conclusion

The staff reviewed the LRA to determine whether any SSCs that should be within the scope of license renewal had not been identified by the applicant. No omissions were identified. In addition, the staff determined whether any components subject to an AMR had not been identified by the applicant. No omissions were identified. The staff concludes that there is reasonable assurance that the applicant has adequately identified the MGAS components within the scope of license renewal, as required by 10 CFR 54.4(a), and those subject to an AMR, as required by 10 CFR 54.21(a)(1).

2.3.4.6 Main Steam System

2.3.4.6.1 Summary of Technical Information in the Application

In LRA Section 2.3.4.6, the applicant described the main steam system, a normally pressurized system designed to deliver steam generated from the RPV system to the main turbine and auxiliary system. The purpose of the main steam system is to provide a primary containment and RCPB function; it serves as the pressure relief system and steam distribution system. It accomplishes the primary containment and RCPB function with piping and valves to limit radiation release rates from the primary containment below the 10 CFR 100 guidelines. It accomplishes the pressure relief function for the RCPB by way of automatic and manual actuation of relief valves. It also provides manual and automatic emergency depressurization by relief valves supporting the core spray system. Distribution of steam to the main turbine and auxiliary system is accomplished by piping distribution branches in the turbine building.

The main steam system contains safety-related components relied upon to remain functional during and following DBEs. The failure of nonsafety-related SSCs in the main steam system potentially could prevent the satisfactory accomplishment of a safety-related function. In addition, the main steam system performs functions that support fire protection and EQ.

The intended functions within the scope of license renewal include:

* maintains mechanical and structural integrity to prevent spatial interactions that could cause failure of safety-related SSCs (includes the required structural support when the nonsafety-related leakage boundary piping is also attached to safety-related piping)

- provides mechanical closure

- provides pressure-retaining boundary; fission product barrier; containment isolation; or containment, holdup, and plateout from MSIV bypass leakage (main steam system)

- provides flow restriction

In LRA Table 2.3.4.6, the applicant identified the following main steam system component types within the scope of license renewal and subject to an AMR:

- closure bolting
- condensing chamber
- coolers (sample)
- eductor
- expansion joint
- flow element (main steam line)
- gauge snubber
- piping and fittings
- sparger (Y-quencher)
- steam trap
- strainer body
- thermowell
- valve body
- valve body (bypass valves)
- valve body (steam chest)

2.3.4.6.2 Staff Evaluation

The staff reviewed LRA Section 2.3.4.6 and UFSAR Sections 5.2.2, 5.2.6.2, 5.4.4, 5.4.5, 6.3.1.2, 7.3, 10.3, and 15.1.5 using the evaluation methodology of SER Section 2.3. The staff conducted its review in accordance with the guidance of SRP-LR Section 2.3.

In conducting its Tier-2 review of the BOP two-tier review process, the staff evaluated the system functions described in the LRA and UFSAR to verify that the applicant had not omitted from the scope of license renewal any components with intended functions under 10 CFR 54.4(a). The staff then reviewed those components that the applicant had identified as within the scope of license renewal to verify that it had not omitted any passive and long-lived components subject to an AMR in accordance with the requirements of 10 CFR 54.21(a)(1).

2.3.4.6.3 Conclusion

The staff reviewed the LRA to determine whether any SSCs that should be within the scope of license renewal had not been identified by the applicant. No omissions were identified. In addition, the staff determined whether any components subject to an AMR had not been identified by the applicant. No omissions were identified. The staff concludes that there is reasonable assurance that the applicant has adequately identified the main steam system components within the scope of license renewal, as required by 10 CFR 54.4(a), and those subject to an AMR, as required by 10 CFR 54.21(a)(1).

2.3.4.7 Main Turbine and Auxiliary System

2.3.4.7.1 Summary of Technical Information in the Application

In LRA Section 2.3.4.7, the applicant described the main turbine and auxiliary systems (MTAS), the purpose of which is to produce rotational energy from the steam generated in the reactor and to discharge exhaust steam into the main condenser. The system accomplishes the purpose by extracting energy from the reactor steam entering the high-pressure turbine through the main stop valves and control valves. Some of the steam is extracted and sent to the first stage reheater. The remaining steam exhausts to the moisture separators and then to the reheaters. Superheated steam from the reheaters is directed to the low-pressure turbines through the combined reheat intercept/stop valves. From there the steam is exhausted to the main condenser. The main turbine and auxiliary system consists of the following subsystems: main turbine (high-pressure and low-pressure turbine sections), mechanical-hydraulic controls front standard, heater drains, vent and pressure relief, moisture separators, reheaters, turbine lubrication oil, lubrication oil purification and transfer, steam seal, turning gear and lift pumps, exhaust hood spray and turbine hood spray, reheat steam, turbine extraction, turbine bypass and the necessary control and protective devices, and operating and supervisory instrumentation.

The failure of nonsafety-related SSCs in the MTAS potentially could prevent the satisfactory accomplishment of a safety-related function.

The intended functions within the scope of license renewal include:

* maintains mechanical and structural integrity to prevent spatial interactions that could cause failure of safety-related SSCs (includes the required structural support when the nonsafety-related leakage boundary piping is also attached to safety-related piping)
* provides mechanical closure

In LRA Table 2.3.4.7, the applicant identified the following MTAS component types within the scope of license renewal and subject to an AMR:

* accumulator
* closure bolting
* coolers
* expansion joint
* filter housing
* flexible hose
* flow element
* heat exchangers
* piping and fittings
* pump casing
* restricting orifice
* sight glasses
* steam trap
* strainer body
* tanks
* thermowell

- turbine casing
- valve body

2.3.4.7.2 Staff Evaluation

The staff reviewed LRA Section 2.3.4.7 and UFSAR Sections 3.5, 7.7.1.5, 10.1, 10.2, 10.3, 10.4, 15.1, and 15.2 using the evaluation methodology of SER Section 2.3. The staff conducted its review in accordance with the guidance of SRP-LR Section 2.3.

In conducting its Tier-1 review of the BOP two-tier review process, the staff evaluated the system functions described in the LRA and UFSAR to verify that the applicant had not omitted from the scope of license renewal any components with intended functions under 10 CFR 54.4(a). The staff then reviewed those components that the applicant had identified as within the scope of license renewal to verify that it had not omitted any passive and long-lived components subject to an AMR in accordance with the requirements of 10 CFR 54.21(a)(1).

2.3.4.7.3 Conclusion

The staff reviewed the LRA to determine whether any SSCs that should be within the scope of license renewal had not been identified by the applicant. No omissions were identified. In addition, the staff determined whether any components subject to an AMR had not been identified by the applicant. No omissions were identified. The staff concludes that there is reasonable assurance that the applicant has adequately identified the MTAS components within the scope of license renewal, as required by 10 CFR 54.4(a), and those subject to an AMR, as required by 10 CFR 54.21(a)(1).

2.4 Scoping and Screening Results: Structures

This section documents the staff's review of the applicant's scoping and screening results for structures. Specifically, this section discusses the following structures and commodity groups:

- primary containment
- reactor building
- chlorination facility
- condensate transfer building
- dilution structure
- emergency diesel generator building
- exhaust tunnel
- fire pond dam
- fire pumphouses
- heating boiler house
- intake structure and canal (ultimate heat sink)
- miscellaneous yard structures
- new radwaste building
- office building
- OCGS substation
- turbine building
- ventilation stack
- component supports commodity group
- piping and component insulation commodity group

In accordance with the requirements of 10 CFR 54.21(a)(1), the applicant must list passive, long-lived SCs within the scope of license renewal and subject to an AMR. To verify that the applicant had properly implemented its methodology, the staff focused its review on the implementation results. This approach allowed the staff to confirm that there were no omissions of structures and components that meet the scoping criteria and are subject to an AMR.

Staff Evaluation Methodology. The staff's evaluation of the information in the LRA was the same for all structures. The objective was to determine whether the components and supporting structures for a specific structure or commodity group, that appeared to meet the scoping criteria specified in the Rule, had been identified by the applicant as within the scope of license renewal, in accordance with 10 CFR 54.4. Similarly, the staff evaluated the applicant's screening results to verify that all long-lived, passive components were subject to an AMR in accordance with 10 CFR 54.21(a)(1).

Scoping. For its evaluation, the staff reviewed the applicable LRA sections and associated component drawings, focusing its review on components that had not been identified as within the scope of license renewal. The staff reviewed relevant licensing basis documents, including the UFSAR, for each structure and commodity group to determine whether the applicant had omitted components with intended functions under 10 CFR 54.4(a) from the scope of license renewal. The staff also reviewed the licensing basis documents to determine whether all intended functions under 10 CFR 54.4(a) were specified in the LRA. If omissions were identified, the staff requested additional information to resolve them.

Screening. After completing its review of the scoping results, the staff evaluated the applicant's screening results. For those SCs with intended functions, the staff sought to determine whether (1) the functions are performed with moving parts or a change in configuration or properties or (2) they are subject to replacement based on a qualified life or specified time period, as described in 10 CFR 54.21(a)(1). For those meeting neither of these criteria, the staff sought to confirm that these SCs were subject to an AMR, as required by 10 CFR 54.21(a)(1). If discrepancies were identified, the staff requested additional information to resolve them.

2.4.1 Primary Containment

2.4.1.1 Summary of Technical Information in the Application

In LRA Section 2.4.1, the applicant described the primary containment structure comprised of the primary containment, containment penetrations, and internal structures. The structure is enclosed by the reactor building, which provides secondary containment, structural support, shielding, shelter, and protection to the containment and components housed within against external design basis events. The primary containment is a General Electric (GE) Mark I design and consists of a drywell, a pressure suppression chamber, and a vent system connecting them. It is designed, fabricated, inspected, and tested in accordance with the requirements of Section VIII of the ASME Boiler and Pressure Vessel Code and Code Cases 1270N-5, 1271N and 1272N-5. The containment is a safety-related, seismic Class I structure. The purpose of the primary containment is to accommodate, with a minimum of leakage, pressures and temperatures resulting from the break of any enclosed process pipe to limit the release of radioactive fission products to offsite dose rate values below 10 CFR Part 100 guideline limits. It also provides a source of water for the ECCS and for pressure suppression in a LOCA. The primary containment is penetrated at several locations by piping, instrument lines, ventilation ducts, and electric leads. Internal structures consist of a fill slab, reactor pedestal, biological shield wall and its lateral support, and structural steel. The primary containment and internal structures also provide structural support to the reactor pressure vessel, the reactor coolant systems, and other safety and nonsafety-related SSCs housed within. The biological shield wall has the added function of radiation shielding to maintain drywell environment within equipment qualification parameters.

In a letter dated December 3, 2006, the applicant provided information concerning the addition of a moisture barrier that was added to the junction of the curb above the fill slab, the drywell shell, and the inside of two trenches, which were excavated on the drywell floor. The applicant also added the moisture barrier to Table 2.4.1.

The primary containment structure contains safety-related components relied upon to remain functional during and following DBEs. The failure of nonsafety-related SSCs in the primary containment structure potentially could prevent the satisfactory accomplishment of a safety-related function. In addition, the primary containment structure performs functions that support fire protection, ATWS, and EQ.

The intended functions within the scope of license renewal include:

- provides enclosure, shelter, or protection for in-scope equipment (including shielding)
- provides HELB shielding
- maintains mechanical and structural integrity to prevent spatial interactions that could cause failure of safety-related SSCs (includes the required structural support when the nonsafety-related leakage boundary piping is also attached to safety-related piping)
- provides pressure-retaining boundary; fission product barrier; containment isolation; or containment, holdup and plateout (main steam system)
- provides shielding against radiation
- provides structural support or structural integrity to preclude nonsafety-related component interactions that could prevent satisfactory accomplishment of a safety-related function

In LRA Table 2.4.1, the applicant identified the following primary containment structure component types within the scope of license renewal and subject to an AMR:

- access hatch covers
- beam seats
- biological shield wall - concrete
- biological shield wall - lateral support
- biological shield wall - liner plate
- biological shield wall - structural steel
- cable tray
- class MC pressure retaining bolting
- concrete embedment
- conduits
- downcomers
- drywell head

- drywell penetration bellows
- drywell penetration sleeves
- drywell shell
- drywell support skirt
- liner (sump)
- locks, hinges, and closure mechanisms
- miscellaneous steel (catwalks, handrails, ladders, platforms, grating, and associated supports)
- panels and enclosures
- penetration closure plates and caps (spare penetrations)
- personnel airlock and equipment hatch
- reactor pedestal
- reinforced concrete floor slab (fill slab)
- seals, gaskets, and o-rings
- shielding blocks and plates
- structural bolting
- structural steel (radial beams, posts, bracing, plate, connections, etc.)
- suppression chamber penetrations
- suppression chamber ring girders
- suppression chamber shell
- suppression chamber shell hoop straps
- thermowells
- vent header
- vent header deflector
- vent jet deflectors
- vent line bellows
- vent line

In a letter dated December 3, 2006, the applicant provided information about the addition of a moisture barrier to the junction of the curb above the fill slab, the drywell shell, and the two trenches excavated on the drywell floor. The applicant also added the moisture barrier to Table 2.4.1.

2.4.1.2 Staff Evaluation

The staff reviewed LRA Section 2.4.1 using the evaluation methodology of SER Section 2.4. The staff conducted its review in accordance with the guidance of SRP-LR Section 2.4, "Scoping and Screening Results: Structures."

In conducting its review, the staff evaluated the structural component functions described in the LRA and UFSAR to verify that the applicant had not omitted from the scope of license renewal any components with intended functions under 10 CFR 54.4(a). The staff then reviewed those components that the applicant had identified as within the scope of license renewal to verify that it had not omitted any passive and long-lived components subject to an AMR in accordance with the requirements of 10 CFR 54.21(a)(1).

The staff's review of LRA Section 2.4.1 identified areas in which additional information was necessary to complete the review of the applicant's scoping and screening results. The applicant responded to the staff's RAIs as discussed below.

In RAI 2.4.1-1 dated March 20, 2006, the staff noted that LRA Table 2.4.1 indicates that drywell seismic support and anchorages are not within the scope of license renewal though relied upon for drywell stability. A component type, "Biological Shield Wall - Lateral Support," is in the table. The staff requested that the applicant justify not including the drywell seismic lateral supports within the scope of license renewal

In its response dated April 18, 2006, the applicant stated that the drywell seismic lateral supports are within the scope of license renewal and subject to AMR. The lateral supports are not specifically identified by name they are included in ASME Class MC component supports and evaluated with the "component supports" commodity group in LRA Section 2.4.18. Their AMR is presented in LRA Table 3.5.2.1.18.

The staff's review of LRA Table 3.5.2.1-18 indicates that the seismic lateral supports are not explicitly included. However, from the first sentence of the response, the staff considers the supports included under the component type "supports for ASME Class MC components." Their aging will be managed by the ASME Section XI, Subsection IWF Program. From the response, the staff finds that the seismic lateral supports are included within the scope of license renewal. The staff's concern described in RAI 2.4.2-1 is resolved.

In RAI 2.4.1-2 dated March 20, 2006, the staff stated that LRA Tables 2.4.1 and 2.4.2 do not include refueling cavity seal components within the scope of license renewal though the plant has experienced significant corrosion (as described in item number 3.5.2.2-4 of LRA Section 3.5.2.2) of the drywell from leakage from the seal. The staff requested that the applicant include the seal within the scope of license renewal or justify not including it.

In its response dated April 18, 2006, the applicant explained that LRA Section 2.4.2 describes the refueling cavity seals and refers to them as refueling bellows, which are classified as nonsafety-related and perform their design function only when the plant is shut down for refueling. Moreover, the applicant noted that refueling bellows are not credited in the CLB for DBEs or accidents, that their failure would not impact a safety function, and that scoping had determined that they perform no 10 CFR 54.4(a) intended function; thus, they are not included in LRA Table 2.4.2.

The applicant also stated that the cavity seals are addressed in RAI 4.7.2-3. In its response to RAI 4.7.2-3 dated April 7, 2006, the applicant provided the following information:

> The refueling seals at Oyster Creek consist of stainless steel bellows. In the mid to late 1980's, GPU conducted extensive visual and NDE inspections to determine the source of water intrusion into the seismic gap between the drywell concrete shield wall and the drywell shell, and its accumulation in the sand bed region. The inspections concluded that the refueling bellows (seals) were not the source of water leakage. The bellows were repeatedly tested using helium (external) and air (internal) without any indication of leakage. Furthermore, any minor leakage from the refueling bellows would be collected in a concrete trough below the bellows. The concrete trough is equipped with a drain line that would direct any leakage to the reactor building equipment drain tank and prevent it from entering the seismic gap (see Figures 1 and 2). The drain line has been checked before refueling outages to confirm it is not blocked.

> The only other seal is the gasket for the reactor cavity seal trough drain line. This

gasket was replaced after the tests showed that it was leaking (see Figure 2). However the gasket leak was ruled out as the primary source of water observed in the sand bed drains because there is no clear leakage path to the seismic gap. Minor gasket leakage would be collected in the concrete trough below the gasket and would be removed by the drain line similar to leaks from the refueling bellows.

Additional visual and NDE (dye penetrant) inspections on the reactor cavity stainless steel liner identified a significant number of cracks, some of which were through wall cracks. Engineering analysis concluded that the cracks were most probably caused by mechanical impact or thermal fatigue and not intergranular stress corrosion cracking (IGSCC). These cracks were determined to be the source of refueling water that passes through the seismic gap. To prevent leakage through the cracks, GPU installed an adhesive type stainless steel tape to bridge any observed large cracks, and subsequently applied the strippable coating. This repair successfully greatly reduced leakage and is implemented every refueling outage while the reactor cavity is flooded. Oyster Creek is currently committed to monitor the sand bed region drains for water leakage. A review of plant documentation did not provide objective evidence that the commitment has been implemented since 1998. Issue Report #348545 was issued in accordance with the Oyster Creek corrective action process to document the lapse in implementing the commitment and to reinforce strict compliance with commitment implementation in the future, including during the period of extended operation.

In addition to the commitment to monitor the sand bed region drains and the reactor cavity concrete trough drains for water leakage (see Figures 1 and 2), Oyster Creek is committed to performing augmented inspections of the drywell in accordance with ASME Section XI, Subsection IWE during the period of extended operation. These inspections consist of periodic UT examinations of the upper region of the drywell and visual examinations of the protective coating on the exterior of the drywell shell in the sand bed region. The visual inspection of the coating will be supplemented by UT measurements from inside the drywell once prior to entering the period of extended operation, and every 10 years thereafter during the period of extended operation.

The staff finds that the refueling seal (bellows) is nonsafety-related. However, its malfunction (including that of the trough drains) could jeopardize the integrity of the drywell shell and, pursuant to 10 CFR 54.4(a)(2), the seal and its components (e.g., drains) must be included within the scope of license renewal.

In addition, the response indicated that the stainless steel liner had cracked at several places. However, from the discussion in LRA Section 2.4.2, the staff understood that the refueling cavity floors and walls (including the stainless steel liner) are within the scope of license renewal and that degradation of these structures and components is managed by the Structures Monitoring Program. Therefore, the staff requested that the applicant include the refueling seal and associated components within the scope of license renewal.

In its supplemental response dated July 7, 2006, the applicant revised Commitment No. 27 to include the following statement:

The reactor cavity concrete trough drain will be verified to be clear from blockage once per refueling cycle. Any identified issues will be addressed via the corrective action process.

The staff believes that in a failure of the bellows or the seal gasket water will accumulate in the trough and, if the drainage from the trough is blocked, water from the trough is likely to get into the air gap between the drywell and the shield concrete. As the applicant committed to monitor the trough drains during each refueling cycle, the potential for water to get into the air gap is reduced substantially. With the applicant's commitment (Commitment No. 27) to utilize the strippable coating during each refueling cycle, the staff finds the applicant's response acceptable. The staff's concern described in RAI 4.7.2-3 is resolved.

2.4.1.3 Conclusion

The staff reviewed the LRA, related structural components, and the RAI responses to determine whether any SSCs that should be within the scope of license renewal had not been identified by the applicant. No omissions were identified. In addition, the staff determined whether any components subject to an AMR had not been identified by the applicant. No omissions were identified. The staff concludes that the applicant has adequately identified the primary containment structure components within the scope of license renewal, as required by 10 CFR 54.4(a), and those subject to an AMR, as required by 10 CFR 54.21(a)(1).

2.4.2 Reactor Building

2.4.2.1 Summary of Technical Information in the Application

In LRA Section 2.4.2, the applicant described the reactor building as designed to completely enclose both the reactor pressure vessel and the primary containment structure, providing a secondary containment. The building is designed to seismic Class I criteria and constructed of reinforced concrete to the refueling floor level. Above the refueling floor, the structure is steel framework with insulated, corrosion-resistant metal siding. The purpose of the reactor building is to provide secondary containment when the primary containment is in service and to provide primary containment during reactor refueling and maintenance operations when the primary containment system is open. The primary objective of the building is to minimize ground level release of airborne radioactive materials and to provide for controlled, elevated release through the ventilation stack to the atmosphere under accident conditions. During normal plant operation, a slight negative pressure is maintained in the building by the reactor building heating and ventilation system so that any leakage is into the building. In an emergency condition, the reactor building heating and ventilation system is isolated and the SGTS serves the building.

The reactor building contains safety-related components relied upon to remain functional during and following DBEs. The failure of nonsafety-related SSCs in the reactor building potentially could prevent the satisfactory accomplishment of a safety-related function. In addition, the reactor building performs functions that support fire protection, ATWS, SBO, and EQ.

The intended functions within the scope of license renewal include:

- provides spray shield or curbs for directing flow
- provides enclosure, shelter, or protection for in-scope equipment (including shielding)

- provides flood protection barrier (internal and external flood event)
- provides HELB shielding
- maintains mechanical and structural integrity to prevent spatial interactions that could cause failure of safety-related SSCs (includes the required structural support when the nonsafety-related leakage boundary piping is also attached to safety-related piping)
- provides missile barrier (internal or external)
- provides pipe whip restraint
- provides shielding against radiation
- provides structural support or structural integrity to preclude nonsafety-related component interactions that could prevent satisfactory accomplishment of a safety-related function
- provides an essentially water leak-tight boundary

In LRA Table 2.4.2, the applicant identified the following reactor building component types within the scope of license renewal and subject to an AMR:

- cable tray
- concrete embedments
- conduits
- curb
- door
- equipment foundation
- fuel pool gates
- fuel pool liner
- fuel pool skimmer surge tank liner
- hatch plugs
- instrument racks
- liner (sump)
- masonry block walls
- metal deck (roof)
- metal siding
- miscellaneous steel: catwalks, handrails, ladders, platforms, grating
- panels and enclosures
- penetration seals
- pipe whip restraints
- reinforced concrete foundation
- reinforced concrete walls (above and below grade)
- reinforced concrete: beams, columns
- reinforced concrete: walls, slabs, drywell shield wall
- roofing
- scuppers: pipe sleeve, flashing, bolts
- seals
- spray shields
- structural bolts
- structural steel: beams, columns, girders, plates, bracing, trusses
- tube tray

2.4.2.2 Staff Evaluation

The staff reviewed LRA Section 2.4.2 using the evaluation methodology of SER Section 2.4. The staff conducted its review in accordance with the guidance of SRP-LR Section 2.4.

In conducting its review, the staff evaluated the structural component functions described in the LRA and UFSAR to verify that the applicant had not omitted from the scope of license renewal any components with intended functions under 10 CFR 54.4(a). The staff then reviewed those components that the applicant had identified as within the scope of license renewal to verify that it had not omitted any passive and long-lived components subject to an AMR in accordance with the requirements of 10 CFR 54.21(a)(1).

The staff's review of LRA Section 2.4.2 identified an area in which additional information was necessary to complete the evaluation of the applicant's scoping and screening results. The applicant responded to the staff's RAIs as discussed below.

In RAI 2.4.2-1 dated March 20, 2006, the staff stated that structural seals are within the boundary of evaluation, as stated in LRA Section 2.4.8, but that the applicant had not explained what they were. The staff requested that the applicant identify all structural seals in the reactor building.

In its response dated April 18, 2006, the applicant stated that component type structural seals or "seals" designates seals other than those specifically used to fill penetrations. For the reactor building, these seals consist of elastomers used as sealant for the superstructure metal siding, flood door seals, HELB door seals, secondary containment door seals, and seals in expansion joints of exterior concrete walls of the building. The seals perform a leakage boundary intended function as designated in LRA Table 3.5.2.1.2.

The applicant clarified what the seals were and listed all the seals in the reactor building. The staff's concern described in RAI 2.4.2-1 is resolved.

2.4.2.3 Conclusion

The staff reviewed the LRA, related structural components, and the RAI response to determine whether any SSCs that should be within the scope of license renewal had not been identified by the applicant. No omissions were identified. In addition, the staff determined whether any components subject to an AMR had not been identified by the applicant. No omissions were identified. The staff concludes that there is reasonable assurance that the applicant has adequately identified the reactor building components within the scope of license renewal, as required by 10 CFR 54.4(a), and those subject to an AMR, as required by 10 CFR 54.21(a)(1).

2.4.3 Chlorination Facility

2.4.3.1 Summary of Technical Information in the Application

In LRA Section 2.4.3, the applicant described the chlorination facility consisting of the chlorination building, spill retention pit, foundation pad for hypochlorite storage tanks, and foundation pads required to support chlorination components. The purpose of the chlorination facility is to provide structural support, shelter, and protection to chlorination, and a 480V motor control center which provides power to the condensate transfer pumps located in the adjacent condensate transfer building. The building is a single-story steel structure with insulated metal siding located west of the reactor building. The base slab is founded on reinforced concrete piers supported from the circulating water tunnel located directly below the building. Foundations for the hypochlorite tanks and other equipment are reinforced concrete pads founded on a common slab with the building and piers supported from the circulating water tunnel. The facility is an nonsafety-related, seismic Class II structure.

The chlorination facility performs functions that support fire protection.

The intended functions within the scope of license renewal include:

- provides enclosure, shelter, or protection for in-scope equipment (including shielding)
- provides structural support or structural integrity to preclude nonsafety-related component interactions that could prevent satisfactory accomplishment of a safety-related function

In LRA Table 2.4.3, the applicant identified the following chlorination facility component types within the scope of license renewal and subject to an AMR:

- conduits
- door
- metal deck
- metal siding
- panels and enclosures
- reinforced concrete foundation
- seals
- structural bolts
- structural steel: beams, columns

2.4.3.2 Staff Evaluation

The staff reviewed LRA Section 2.4.3 using the evaluation methodology of SER Section 2.4. The staff conducted its review in accordance with the guidance of SRP-LR Section 2.4.

In conducting its review, the staff evaluated the structural component functions described in the LRA and UFSAR to verify that the applicant had not omitted from the scope of license renewal any components with intended functions under 10 CFR 54.4(a). The staff then reviewed those components that the applicant had identified as within the scope of license renewal to verify that it had not omitted any passive and long-lived components subject to an AMR in accordance with the requirements of 10 CFR 54.21(a)(1).

2.4.3.3 Conclusion

The staff reviewed the LRA and related structural components to determine whether any SSCs that should be within the scope of license renewal had not been identified by the applicant. No omissions were identified. In addition, the staff determined whether any components subject to an AMR had not been identified by the applicant. No omissions were identified. The staff concludes that there is reasonable assurance that the applicant has adequately identified the chlorination facility components within the scope of license renewal, as required by 10 CFR 54.4(a), and those subject to an AMR, as required by 10 CFR 54.21(a)(1).

2.4.4 Condensate Transfer Building

2.4.4.1 Summary of Technical Information in the Application

In LRA Section 2.4.4, the applicant described the condensate transfer building as a single-story steel structure with metal siding located west of the reactor building. The purpose of the condensate transfer building is to provide structural support, shelter, and protection for the condensate transfer pumps, demineralized water transfer pumps, and service water booster pump. The base slab is founded on reinforced concrete piers supported from the circulating water tunnel located directly below the building. A half-ton hoist is incorporated in the design of the structure to facilitate removal and maintenance of equipment. The structure is classified as nonsafety-related, seismic Class II.

The condensate transfer building performs functions that support fire protection.

The intended functions within the scope of license renewal include:

- provides enclosure, shelter, or protection for in-scope equipment (including shielding)
- provides structural support or structural integrity to preclude nonsafety-related component interactions that could prevent satisfactory accomplishment of a safety-related function

In LRA Table 2.4.4, the applicant identified the following condensate transfer building component types within the scope of license renewal and subject to an AMR:

- conduits
- door
- equipment foundation
- metal deck
- metal siding
- panels and enclosures
- reinforced concrete foundation (includes piers)
- seals
- structural bolts
- structural steel: beams, columns

2.4.4.2 Staff Evaluation

The staff reviewed LRA Section 2.4.4 using the evaluation methodology of SER Section 2.4. The staff conducted its review in accordance with the guidance of SRP-LR Section 2.4.

In conducting its review, the staff evaluated the structural component functions described in the LRA and UFSAR to verify that the applicant had not omitted from the scope of license renewal any components with intended functions under 10 CFR 54.4(a). The staff then reviewed those components that the applicant had identified as within the scope of license renewal to verify that it had not omitted any passive and long-lived components subject to an AMR in accordance with

2-142

the requirements of 10 CFR 54.21(a)(1).

2.4.4.3 Conclusion

The staff reviewed the LRA and related structural components to determine whether any SSCs that should be within the scope of license renewal had not been identified by the applicant. No omissions were identified. In addition, the staff determined whether any components subject to an AMR had not been identified by the applicant. No omissions were identified. The staff concludes that there is reasonable assurance that the applicant has adequately identified the condensate transfer building components within the scope of license renewal, as required by 10 CFR 54.4(a), and those subject to an AMR, as required by 10 CFR 54.21(a)(1).

2.4.5 Dilution Structure

2.4.5.1 Summary of Technical Information in the Application

In LRA Section 2.4.5, the applicant described the dilution structure located west of the reactor building on the west bank of the intake canal. The purpose of the dilution structure is to house the dilution system and its supporting systems. The structure provides physical support, shelter, and protection to nonsafety-related components designed to divert water from the intake canal to the discharge canal for thermal dilution. Additionally, the structure in conjunction with earthen dikes forms the intake canal boundary and separates it from the discharge canal. The structure is of reinforced concrete, approximate 83 feet long and divided into three bays, each with two trash racks and one dilution pump. The three dilution pumps discharge into a common reinforced concrete tunnel that delivers dilution water from the intake canal to the discharge canal. Sheet metal and wooden enclosures located on the top slab of the structure at grade level provide shelter for pump motors and other dilution system components. The foundation for the structure consists of a reinforced concrete slab, with shear keys, founded on soil 30 foot below grade level. Stop logs are incorporated into the structure's design to isolate each bay from the intake canal. The structure is classified as nonsafety-related, seismic Class II.

The failure of nonsafety-related SSCs in the dilution structure potentially could prevent the satisfactory accomplishment of a safety-related function.

The intended functions within the scope of license renewal include:

- provides structural support or structural integrity to preclude nonsafety-related component interactions that could prevent satisfactory accomplishment of a safety-related function
- provides an essentially water leak-tight boundary

In LRA Table 2.4.5, the applicant identified the following dilution structure component types within the scope of license renewal and subject to an AMR:

- reinforced concrete foundation
- reinforced concrete walls

2.4.5.2 Staff Evaluation

The staff reviewed LRA Section 2.4.5 using the evaluation methodology of SER Section 2.4. The staff conducted its review in accordance with the guidance of SRP-LR Section 2.4.

In conducting its review, the staff evaluated the structural component functions described in the LRA and UFSAR to verify that the applicant had not omitted from the scope of license renewal any components with intended functions under 10 CFR 54.4(a). The staff then reviewed those components that the applicant had identified as within the scope of license renewal to verify that it had not omitted any passive and long-lived components subject to an AMR in accordance with the requirements of 10 CFR 54.21(a)(1).

2.4.5.3 Conclusion

The staff reviewed the LRA and related structural components to determine whether any SSCs that should be within the scope of license renewal had not been identified by the applicant. No omissions were identified. In addition, the staff determined whether any components subject to an AMR had not been identified by the applicant. No omissions were identified. The staff concludes that there is reasonable assurance that the applicant has adequately identified the dilution structure components within the scope of license renewal, as required by 10 CFR 54.4(a), and those subject to an AMR, as required by 10 CFR 54.21(a)(1).

2.4.6 Emergency Diesel Generator Building

2.4.6.1 Summary of Technical Information in the Application

In LRA Section 2.4.6, the applicant described the EDG building as a single-story structure located southwest of the reactor building. The purpose of the EDG building is to provide support, shelter, and protection for each EDG, the diesel oil storage tank, and components of the fuel transfer system. The reinforced concrete structure consists of two compartments, one for each EDG, and an appendage vault to the building containing the diesel oil storage tank. Personnel entrances to the building have reinforced concrete labyrinth walls for missile protection and a 6-inch high curb for flood protection. The building foundation is reinforced concrete slab on grade. The building is classified as safety-related, designed to seismic Class I. Each EDG is also housed in a metal enclosure which provides protection against rain, snow, and dust that may enter the building through the air intake and exhaust openings on the roof. The building also houses and supports such nonsafety-related components as grating, lighting conduit, and electrical enclosures.

The EDG building contains safety-related components relied upon to remain functional during and following DBEs. The failure of nonsafety-related SSCs in the EDG building potentially could prevent the satisfactory accomplishment of a safety-related function. In addition, the EDG building performs functions that support fire protection and SBO.

The intended functions within the scope of license renewal include:

* provides spray shield or curbs for directing flow

* provides enclosure, shelter, or protection for in-scope equipment (including shielding)

* provides flood protection barrier (internal and external flood event)

2-144

- provides missile barrier (internal or external)
- provides structural support or structural integrity to preclude nonsafety-related component interactions that could prevent satisfactory accomplishment of a safety-related function

In LRA Table 2.4.6, the applicant identified the following EDG building component types within the scope of license renewal and subject to an AMR:

- concrete embedments
- conduits
- curb
- EDG enclosure
- miscellaneous steel (catwalks, handrails, ladders, platforms, grating, and associated supports)
- panels and enclosures
- reinforced concrete foundation
- reinforced concrete walls, slabs (includes removable roof slab)
- structural bolts
- structural steel (plate)

2.4.6.2 Staff Evaluation

The staff reviewed LRA Section 2.4.6 using the evaluation methodology of SER Section 2.4. The staff conducted its review in accordance with the guidance of SRP-LR Section 2.4.

In conducting its review, the staff evaluated the structural component functions described in the LRA and UFSAR to verify that the applicant had not omitted from the scope of license renewal any components with intended functions under 10 CFR 54.4(a). The staff then reviewed those components that the applicant had identified as within the scope of license renewal to verify that it had not omitted any passive and long-lived components subject to an AMR in accordance with the requirements of 10 CFR 54.21(a)(1).

2.4.6.3 Conclusion

The staff reviewed the LRA and related structural components to determine whether any SSCs that should be within the scope of license renewal had not been identified by the applicant. No omissions were identified. In addition, the staff determined whether any components subject to an AMR had not been identified by the applicant. No omissions were identified. The staff concludes that there is reasonable assurance that the applicant has adequately identified the EDG building components within the scope of license renewal, as required by 10 CFR 54.4(a), and those subject to an AMR, as required by 10 CFR 54.21(a)(1).

2.4.7 Exhaust Tunnel

2.4.7.1 Summary of Technical Information in the Application

In LRA Section 2.4.7, the applicant described the exhaust tunnel, which consists of an underground reinforced concrete box that connects the ventilation stack, the reactor building, turbine building, and the old radwaste building. The purpose of the exhaust tunnel is to provide structural support, shelter, and protection for the SGTS components and ductwork and for 4160V AC and 480V AC electrical cables. It also provides structural support, shelter, and protection for nonsafety-related system piping and ductwork routed within the tunnel. The tunnel houses major components of the SGTS with the exception of the exhaust fans and outlet valves. The tunnel also routes reactor building ventilation, turbine building ventilation, and old radwaste building ventilation exhaust ductwork to the ventilation stack as well as process piping and drain lines routed between the buildings. Also routed through the tunnel are 4160V AC system cables, which feed core spray pumps, and 480V AC system power to the SGTS components. In addition, the tunnel contains heating steam piping routed from the heating boiler house to the buildings. The exhaust tunnel is classified as an nonsafety-related, seismic Class II structure.

The failure of nonsafety-related SSCs in the exhaust tunnel potentially could prevent the satisfactory accomplishment of a safety-related function. The exhaust tunnel also performs functions that support fire protection and EQ.

The intended functions within the scope of license renewal include:

- provides spray shield or curbs for directing flow

- provides enclosure, shelter, or protection for in-scope equipment (including shielding)

- maintains mechanical and structural integrity to prevent spatial interactions that could cause failure of safety-related SSCs (includes the required structural support when the nonsafety-related leakage boundary piping is also attached to safety-related piping)

- provides structural support or structural integrity to preclude nonsafety-related component interactions that could prevent satisfactory accomplishment of a safety-related function

In LRA Table 2.4.7, the applicant identified the following exhaust tunnel component types within the scope of license renewal and subject to an AMR:

- concrete embedments
- conduits
- curb
- door
- hatch cover
- masonry block walls
- panels and enclosures
- penetration seals
- reinforced concrete slabs
- seals (gap)
- walls

2.4.7.2 Staff Evaluation

The staff reviewed LRA Section 2.4.7 using the evaluation methodology of SER Section 2.4. The staff conducted its review in accordance with the guidance of SRP-LR Section 2.4.

In conducting its review, the staff evaluated the structural component functions described in the LRA and UFSAR to verify that the applicant had not omitted from the scope of license renewal any components with intended functions under 10 CFR 54.4(a). The staff then reviewed those components that the applicant had identified as within the scope of license renewal to verify that it had not omitted any passive and long-lived components subject to an AMR in accordance with the requirements of 10 CFR 54.21(a)(1).

2.4.7.3 Conclusion

The staff reviewed the LRA and related structural components to determine whether any SSCs that should be within the scope of license renewal had not been identified by the applicant. No omissions were identified. In addition, the staff determined whether any components subject to an AMR had not been identified by the applicant. No omissions were identified. The staff concludes that there is reasonable assurance that the applicant has adequately identified the exhaust tunnel components within the scope of license renewal, as required by 10 CFR 54.4(a), and those subject to an AMR, as required by 10 CFR 54.21(a)(1).

2.4.8 Fire Pond Dam

2.4.8.1 Summary of Technical Information in the Application

In LRA Section 2.4.8, the applicant described the fire pond dam constructed across the OCGS stream outside the protected area and approximately 1/4 mile from the reactor building. The purpose of the fire pond dam is to contain fresh water for use in the fire protection system. Water from the pond is supplied to the fire protection system by two pumps housed in the fresh water pump house adjacent to the dam. The dam is 130 feet long and consists of two parallel lines of tongue and grooved wood sheeting 5 feet apart and driven into the channel bottom. The area between the upstream and downstream sheeting is lined with a 4-inch reinforced concrete slab which forms a shallow open channel that directs water flow to a 45-foot wide stream spillway. Rip-rap is placed downstream of the spillway to protect the stream from erosion. The pond formed by the dam covers over 6 acres of land and has a volume equivalent to 7.2 million gallons of water. The dam, classified safety Class III, is subject to State of New Jersey Department of Environmental Protection and Energy dam safety regulations.

The fire pond dam performs functions that support fire protection.

The intended function, within the scope of license renewal, is to provide an essentially water leak-tight boundary.

In LRA Table 2.4.8, the applicant identified the component type fire pond dam structure as within the scope of license renewal and subject to an AMR.

2.4.8.2 Staff Evaluation

The staff reviewed LRA Section 2.4.8 using the evaluation methodology of SER Section 2.4. The staff conducted its review in accordance with the guidance of SRP-LR Section 2.4.

In conducting its review, the staff evaluated the structural component functions described in the LRA and UFSAR to verify that the applicant had not omitted from the scope of license renewal any components with intended functions under 10 CFR 54.4(a). The staff then reviewed those components that the applicant had identified as within the scope of license renewal to verify that it had not omitted any passive and long-lived components subject to an AMR in accordance with the requirements of 10 CFR 54.21(a)(1).

The staff's review of LRA Section 2.4.8 identified an area in which additional information was necessary to complete the review of the applicant's scoping and screening results. The applicant responded to the staff's RAIs as discussed below.

In RAI 2.4.8-1 dated March 30, 2006, the staff noted that in LRA Section 2.4.8 the fire pond dam is classified as safety Class III. The staff requested that the applicant identify in the LRA or UFSAR the definition of safety Class III. If the definition was not in the LRA or UFSAR, the staff requested that the applicant provide a definition.

In its response dated April 18, 2006, the applicant stated that the fire pond dam classification is related to the hazard potential of property damage or loss of life if the dam failed, not to nuclear safety, and is not defined in the UFSAR. The term is not defined in the LRA because it does not affect scoping, screening, and aging management of the dam. The fire pond dam is within the scope of license renewal under 10 CFR 54.4(a)(3) and is relied upon in the safety analyses and plant evaluations to perform a function for compliance with NRC fire protection regulations. As described in LRA Section 2.4.8, the dam is classified safety Class III and subject to State of New Jersey Department of Environment Protection and Energy dam safety regulations. The safety Class III classification is assigned by the State of New Jersey to dams the failure of which would not cause loss of life or significant property damage. This classification is synonymous with the "low-hazard potential" assigned to dams in the Federal Emergency Management Agency guidelines for dam safety.

The staff concludes that the applicant's response had provided an adequate explanation of safety Class III. The staff's concern described in RAI 2.4.8-1 is resolved.

2.4.8.3 Conclusion

The staff reviewed the LRA, related structural components, and the RAI response to determine whether any SSCs that should be within the scope of license renewal had not been identified by the applicant. No omissions were identified. In addition, the staff determined whether any components subject to an AMR had not been identified by the applicant. No omissions were identified. The staff concludes that there is reasonable assurance that the applicant has adequately identified the fire pond dam components within the scope of license renewal, as required by 10 CFR 54.4(a), and those subject to an AMR, as required by 10 CFR 54.21(a)(1).

2.4.9 Fire Pumphouses

2.4.9.1 Summary of Technical Information in the Application

In LRA Section 2.4.9, the applicant described the fire pumphouses, the purpose of which is to provide structural support, shelter, and protection for fire protection system components and for components supporting the intended function of the system. The fire pumphouses are comprised of the fresh water pumphouse and the redundant fire protection pumphouse. The fresh water pumphouse is located west of the reactor building outside the protected area. It consists of a prefabricated sheet metal enclosure, an intake reinforced concrete structure, and foundations for two fuel oil tanks. The intake structure is divided into three separate pump intake bays, one for each of the two vertical centrifugal diesel engine-driven fire pumps and one for two electric pond pumps. The inlet into the bays is protected with trash racks and stationary water screens. The two diesel-driven pumps supply primary fire water, drawn from a pond formed by a small dam, to the fire protection system. The two electric pond pumps maintain fire water system pressure. The pumps, the diesel engines, and their supporting systems are inside the enclosure supported from the roof slab of the intake bays. The fuel oil tanks are outside the enclosure within a diked area and independently supported by reinforced concrete foundations. A monorail outside the enclosure, supported on structural frames, provides the means for cleaning and servicing the stationary water screens. The pumphouse and the tank foundations are classified as nonsafety-related, seismic Class II. The redundant fire protection pumphouse is northwest of the reactor building inside the protected area. It consists of a prefabricated sheet metal enclosure, foundation slab on grade, and foundation for the redundant fire protection water tank. The structure houses a motor-driven electric fire pump and its supporting electrical systems. This pump and its tank constitute an emergency supply when the primary supply is not available. The pumphouse is classified as nonsafety-related, seismic Class II.

The fire pumphouses perform functions that support fire protection.

The intended functions within the scope of license renewal include:

- provides enclosure, shelter, or protection for in-scope equipment (including shielding)
- provides structural support or structural integrity to preclude nonsafety-related component interactions that could prevent satisfactory accomplishment of a safety-related function

In LRA Table 2.4.9, the applicant identified the following fire pumphouses component types within the scope of license renewal and subject to an AMR:

- conduits
- metal deck
- metal siding
- panels and enclosures
- reinforced concrete foundation
- reinforced concrete slab
- reinforced concrete walls
- seals
- structural bolts
- structural steel

2.4.9.2 Staff Evaluation

The staff reviewed LRA Section 2.4.9 using the evaluation methodology of SER Section 2.4. The staff conducted its review in accordance with the guidance of SRP-LR Section 2.4.

In conducting its review, the staff evaluated the structural component functions described in the LRA and UFSAR to verify that the applicant had not omitted from the scope of license renewal any components with intended functions under 10 CFR 54.4(a). The staff then reviewed those components that the applicant had identified as within the scope of license renewal to verify that it had not omitted any passive and long-lived components subject to an AMR in accordance with the requirements of 10 CFR 54.21(a)(1).

The staff's review of LRA Section 2.4.9 identified an area in which additional information was necessary to complete the review of the applicant's scoping and screening results. The applicant responded to the staff's RAIs as discussed below.

In RAI 2.4.9-1 dated March 30, 2006, the staff stated that LRA Section 2.4.9 classifies the pumphouse and the tank foundations as nonsafety-related, seismic Class II. The staff requested that the applicant identify in the LRA or UFSAR the definition of "nonsafety-related, seismic Class II." If the definition was not in the LRA or UFSAR, the staff requested that the applicant provide a definition.

In its response dated April 18, 2006, the applicant stated:

> Seismic classification of structures is defined in UFSAR Section 3.8.3.2, Applicable Codes, Standards and Specifications, and Section 3.8.4.1, Description of the Structures. According to these sections, there are two classes of structures for which earthquake design requirements apply as follows:
>
> - Class I: Structures and equipment whose failure could cause significant release of radioactivity or which are vital to a proper shutdown of the plant and the removal of decay heat.
>
> - Class II: Structures and equipment which are both essential and nonessential to the operation of the station, but which are not essential to a proper shutdown.
>
> The Fire Pumphouses and tank foundations are classified Seismic Class II structures based on UFSAR definition above. For license renewal, the Fire Pumphouses and the tank foundations meet 10 CFR 54.4(a)(3) because they are relied upon in the safety analyses and plant evaluations to perform a function that demonstrates compliance with the Commission's regulations for fire protection (10 CFR 50.48). The pumphouses and the tank foundations do not meet 10 CFR 54.4(a)(1) because they are not safety-related structures that are relied on to remain functional during and following design basis events. The pumphouses and the tank foundations do not meet 10 CFR 54.4(a)(2) because failure of non-safety related portions of the structures would not prevent satisfactory accomplishment of function(s) identified for 10 CFR 54.4(a)(1). The structures are not relied upon in any safety analyses or plant evaluations to perform a function that demonstrates compliance with the Commission's regulation for Environmental Qualification (10 CFR 50.49), ATWS (10 CFR 50.62), or Station Blackout (10 CFR 50.63).

The staff concludes that the applicant's response was acceptable as it clearly defined the seismic Class II pumphouse and the tank foundation. The staff's concern described in RAI 2.4.9-1 is resolved.

2.4.9.3 Conclusion

The staff reviewed the LRA, related structural components, and the RAI response to determine whether any SSCs that should be within the scope of license renewal had not been identified by the applicant. No omissions were identified. In addition, the staff determined whether any components subject to an AMR had not been identified by the applicant. No omissions were identified. The staff concludes that there is reasonable assurance that the applicant has adequately identified the fire pumphouse components within the scope of license renewal, as required by 10 CFR 54.4(a), and those subject to an AMR, as required by 10 CFR 54.21(a)(1).

2.4.10 Heating Boiler House

2.4.10.1 Summary of Technical Information in the Application

In LRA Section 2.4.10, the applicant described the heating boiler house license renewal structure comprised of the old and the new heating boiler house. The purpose of the structures is to house the nonsafety-related heating and process steam system components and supporting systems. Major components housed in the buildings include oil-fired boilers, heating boiler feed pumps, fuel oil pumps, deaerator, chemical tanks and feed pumps, boiler condensate storage tank, and system piping. Each heating boiler house is a single-story steel structure located southeast of the reactor building. The buildings are enclosed with insulated metal siding, roof metal deck, and built-up roofing. Foundations for the structures consist of reinforced concrete isolated footings and a reinforced concrete base slab on grade. The old heating boiler house is adjacent and provides access to the ventilation stack through a double door airlock. It also houses two safety-related electrical load centers, electrical panels and enclosures, a transformer, and electrical conduits required for the operation of the SGTS fans. The new heating boiler house does not house any safety-related SSCs. The two heating boiler houses are classified as nonsafety-related, seismic Class II.

The failure of nonsafety-related SSCs in the old heating boiler house potentially could prevent the satisfactory accomplishment of a safety-related function. The old heating boiler house also performs functions that support fire protection.

The intended functions within the scope of license renewal include:

- provides enclosure, shelter, or protection for in-scope equipment (including shielding)
- provides structural support or structural integrity to preclude nonsafety-related component interactions that could prevent satisfactory accomplishment of a safety-related function

In LRA Table 2.4.10, the applicant identified the following heating boiler house component types within the scope of license renewal and subject to an AMR:

- conduits
- door
- equipment foundation
- metal deck
- metal siding
- panels and enclosures
- reinforced concrete foundation
- removable panel (in siding)
- seals
- structural bolts
- structural steel: beams, columns, girts, bracing, connection plates and angles

2.4.10.2 Staff Evaluation

The staff reviewed LRA Section 2.4.10 using the evaluation methodology of SER Section 2.4. The staff conducted its review in accordance with the guidance of SRP-LR Section 2.4.

In conducting its review, the staff evaluated the structural component functions described in the LRA and UFSAR to verify that the applicant had not omitted from the scope of license renewal any components with intended functions under 10 CFR 54.4(a). The staff then reviewed those components that the applicant had identified as within the scope of license renewal to verify that it had not omitted any passive and long-lived components subject to an AMR in accordance with the requirements of 10 CFR 54.21(a)(1).

2.4.10.3 Conclusion

The staff reviewed the LRA and related structural components to determine whether any SSCs that should be within the scope of license renewal had not been identified by the applicant. No omissions were identified. In addition, the staff determined whether any components subject to an AMR had not been identified by the applicant. No omissions were identified. The staff concludes that there is reasonable assurance that the applicant has adequately identified the heating boiler house components within the scope of license renewal, as required by 10 CFR 54.4(a), and those subject to an AMR, as required by 10 CFR 54.21(a)(1).

2.4.11 Intake Structure and Canal (Ultimate Heat Sink)

2.4.11.1 Summary of Technical Information in the Application

In LRA Section 2.4.11, the applicant described the intake structure and canal (ultimate heat sink). The purpose of the intake structure and canal is to provide seawater to dissipate waste heat from the plant during normal, shutdown, and accident conditions. The intake structure also provides structural support for pumps and components that deliver seawater to the plant. In addition, the structure provides structural support and access to electrical, mechanical, and structural components required to support the function and operation of the CWS, SWS, ESW system, screen wash system, and new radwaste SWS, including sluice gates, stop logs, trash racks, trash cart, traveling water intake screens, platforms, ladders, and stairs. The intake structure is composed of reinforced concrete slabs, beams, and shear walls. The structure is largely buried underground or submerged in seawater. Its foundation is a reinforced concrete mat founded on Cohansey sand with a concrete apron that extends into and below the intake canal. The intake canal draws seawater from Barnegat Bay and conveys it to the intake structure. The canal is 140 feet wide, dredged to 10 feet below mean sea level, and separated from the discharge canal by the dilution pump structure and an earthen dike at the intake structure. The canal banks are lined with asphalt bonded stone for protection against erosion. The canal is the ultimate heat sink, required to provide cooling water for emergency shutdown as well as during normal plant operation.

The intake structure and canal contain safety-related components relied upon to remain functional during and following DBEs. The failure of nonsafety-related SSCs in the intake structure and canal potentially could prevent the satisfactory accomplishment of a safety-related function. In addition, the intake structure and canal perform functions that support fire protection.

The intended functions within the scope of license renewal include:

- provides enclosure, shelter, or protection for in-scope equipment (including shielding)
- provides filtration
- provides structural support or structural integrity to preclude nonsafety-related component interactions that could prevent satisfactory accomplishment of a safety-related function
- provides an essentially water leak-tight boundary

In LRA Table 2.4.11, the applicant identified the following intake structure and canal component types within the scope of license renewal and subject to an AMR:

- conduits
- earthen water control structures (intake canal, embankments)
- reinforced concrete foundation
- reinforced concrete slab
- reinforced concrete walls
- trash racks

2.4.11.2 Staff Evaluation

The staff reviewed LRA Section 2.4.11 using the evaluation methodology of SER Section 2.4. The staff conducted its review in accordance with the guidance of SRP-LR Section 2.4.

In conducting its review, the staff evaluated the structural component functions described in the LRA and UFSAR to verify that the applicant had not omitted from the scope of license renewal any components with intended functions under 10 CFR 54.4(a). The staff then reviewed those components that the applicant had identified as within the scope of license renewal to verify that it had not omitted any passive and long-lived components subject to an AMR in accordance with the requirements of 10 CFR 54.21(a)(1).

2.4.11.3 Conclusion

The staff reviewed the LRA and related structural components to determine whether any SSCs that should be within the scope of license renewal had not been identified by the applicant. No omissions were identified. In addition, the staff determined whether any components subject to an AMR had not been identified by the applicant. No omissions were identified. The staff concludes that there is reasonable assurance that the applicant has adequately identified the intake structure and canal components within the scope of license renewal, as required by 10 CFR 54.4(a), and those subject to an AMR, as required by 10 CFR 54.21(a)(1).

2.4.12 Miscellaneous Yard Structures

2.4.12.1 Summary of Technical Information in the Application

In LRA Section 2.4.12, the applicant described the miscellaneous yard structures comprised of concrete and steel structures throughout the yard area. Concrete structures include foundations for outdoor tanks, SGTS fan pads, material storage area pads, transformer foundations, electrical substation components, transmission towers, electrical bus duct supports, trailers, and lighting poles. Concrete structures also include the SWS seal well, sanitary waste system underground concrete tank, trenches, duct banks, manholes, drainage catch basins, concrete retaining walls, concrete curbs, and concrete dikes. Steel structures are comprised of trailers, transmission towers, component supports in the yard (including supports for offsite power system and SBO components), electrical enclosures, 480V switchgear room ventilation fan platforms, and yard storm drainage piping. The purpose of miscellaneous yard structures is to provide structural support, shelter, and protection for safety-related and nonsafety-related components and commodities, including offsite power, SBO, and components credited for fire protection. The purpose of SWS seal well is to reduce the head requirements of the SWS by providing a siphon discharge and a flow path for the SWS. The purpose of curbs and dikes is to contain fluid spills for controlled release. The curb at the entrance to the emergency diesel generator building prevents water intrusion into the building during high floods. Trailers provide additional office space and house nonsafety-related equipment and components not within the scope of license renewal.

The failure of nonsafety-related SSCs in the miscellaneous yard structures potentially could prevent the satisfactory accomplishment of a safety-related function. The miscellaneous yard structures also perform functions that support fire protection and SBO.

The intended functions within the scope of license renewal include:

- provides enclosure, shelter, or protection for in-scope equipment (including shielding)
- provides flood protection barrier (internal and external flood event)
- provides structural support or structural integrity to preclude nonsafety-related component interactions that could prevent satisfactory accomplishment of a safety-related function
- provides an essentially water leak-tight boundary

In LRA Table 2.4.12, the applicant identified the following miscellaneous yard structures component types within the scope of license renewal and subject to an AMR:

- concrete embedments
- conduits
- curb
- equipment and component foundations (startup, unit substation, and SBO transformers, nitrogen supply, SGTS fans and motors, HVAC components, etc.)
- miscellaneous steel (manhole covers)
- miscellaneous steel (platforms)
- panels and enclosures (startup, unit substation, and SBO transformers)
- reinforced concrete trench, manhole, ductbank
- reinforced concrete walls, slabs (SWS seal well)
- structural bolts

- tank foundations (CST, fire water, CO_2, nitrogen, fuel oil)
- transmission towers

2.4.12.2 Staff Evaluation

The staff reviewed LRA Section 2.4.12 using the evaluation methodology of SER Section 2.4. The staff conducted its review in accordance with the guidance of SRP-LR Section 2.4.

In conducting its review, the staff evaluated the structural component functions described in the LRA and UFSAR to verify that the applicant had not omitted from the scope of license renewal any components with intended functions under 10 CFR 54.4(a). The staff then reviewed those components that the applicant had identified as within the scope of license renewal to verify that it had not omitted any passive and long-lived components subject to an AMR in accordance with the requirements of 10 CFR 54.21(a)(1).

2.4.12.3 Conclusion

The staff reviewed the LRA and related structural components to determine whether any SSCs that should be within the scope of license renewal had not been identified by the applicant. No omissions were identified. In addition, the staff determined whether any components subject to an AMR had not been identified by the applicant. No omissions were identified. The staff concludes that there is reasonable assurance that the applicant has adequately identified the miscellaneous yard structure components within the scope of license renewal, as required by 10 CFR 54.4(a), and those subject to an AMR, as required by 10 CFR 54.21(a)(1).

2.4.13 New Radwaste Building

2.4.13.1 Summary of Technical Information in the Application

In LRA Section 2.4.13, the applicant described the new radwaste building as a three-story structure located northeast of the reactor building. The purpose of the new radwaste building is to house the liquid radwaste system, which is classified as nonsafety-related and designed in accordance with the recommendations of RGs 1.26 and 1.29. The building provides structural support, shelter, and protection for the system components and radiation protection during plant operating conditions. Some elements of the building (walls and slabs) are credited, in the CLB, for retention of liquid radwaste during a safe shutdown earthquake. These elements are designed to seismic Class I criteria and sealed watertight. The seismic Class I boundary is based on the volume required to contain the entire liquid inventory of the radwaste system inside the building, taking into account the effects of non-seismic elements of the building collapsing and displacing some of this liquid. This basis provides assurance that postulated failures of the nonseismic liquid radwaste components within the building will not cause uncontrolled releases of radioactivity in liquid form to the environment. The rest of the building is nonseismic, conventionally designed. The building is rectangular in plan, constructed on a reinforced concrete foundation mat at grade resting on compacted backfill. Steel framing and metal decking support the reinforced concrete floor slabs. Walls required to contain liquid radwaste within the building, in the event of liquid radwaste system components failure, are reinforced concrete. Other walls consist of insulated metal siding or solid concrete block construction.

The failure of nonsafety-related SSCs in the new radwaste building potentially could prevent the satisfactory accomplishment of a safety-related function.

The intended functions within the scope of license renewal include:

- provides structural support or structural integrity to preclude nonsafety-related component interactions that could prevent satisfactory accomplishment of a safety-related function
- provides an essentially water leak-tight boundary

In LRA Table 2.4.13, the applicant identified the following new radwaste building component types within the scope of license renewal and subject to an AMR:

- penetration seals
- reinforced concrete foundation
- reinforced concrete walls (above and below grade)

2.4.13.2 Staff Evaluation

The staff reviewed LRA Section 2.4.13 using the evaluation methodology of SER Section 2.4. The staff conducted its review in accordance with the guidance of SRP-LR Section 2.4.

In conducting its review, the staff evaluated the structural component functions described in the LRA and UFSAR to verify that the applicant had not omitted from the scope of license renewal any components with intended functions under 10 CFR 54.4(a). The staff then reviewed those components that the applicant had identified as within the scope of license renewal to verify that it had not omitted any passive and long-lived components subject to an AMR in accordance with the requirements of 10 CFR 54.21(a)(1).

2.4.13.3 Conclusion

The staff reviewed the LRA and related structural components to determine whether any SSCs that should be within the scope of license renewal had not been identified by the applicant. No omissions were identified. In addition, the staff determined whether any components subject to an AMR had not been identified by the applicant. No omissions were identified. The staff concludes that there is reasonable assurance that the applicant has adequately identified the new radwaste building components within the scope of license renewal, as required by 10 CFR 54.4(a), and those subject to an AMR, as required by 10 CFR 54.21(a)(1).

2.4.14 Office Building

2.4.14.1 Summary of Technical Information in the Application

In LRA Section 2.4.14, the applicant described the office building as a three-story concrete structure between the reactor and turbine buildings. The purpose of the office building is to house and support recirculation pump motor generator sets, emergency switchgear, main station batteries, and their electrical and mechanical supporting systems, including ventilation systems. The building also provides offices for site management and plant support personnel, chemistry laboratory testing equipment, showers, locker rooms, and a secondary access to controlled areas. The building is erected partly on the reactor building and partly on a separate mat foundation slab on grade separated from the reactor building by 1-½ inch gap to allow for differential settlement. The reactor building west wall and the torus area roof slab form the east wall of the office building and its first floor slab, respectively. The building was designed as a seismic Class II as specified in UFSAR Section 3.8.4.

The office building contains safety-related components relied upon to remain functional during and following DBEs. The failure of nonsafety-related SSCs in the office building potentially could prevent the satisfactory accomplishment of a safety-related function. In addition, the office building performs functions that support fire protection.

The intended functions within the scope of license renewal include:

- provides spray shield or curbs for directing flow
- provides enclosure, shelter, or protection for in-scope equipment (including shielding)
- provides structural support or structural integrity to preclude nonsafety-related component interactions that could prevent satisfactory accomplishment of a safety-related function

In LRA Table 2.4.14, the applicant identified the following office building component types within the scope of license renewal and subject to an AMR:

- cable tray
- concrete embedments
- conduits
- curb
- masonry block walls
- panels and enclosures
- reinforced concrete foundation
- reinforced concrete walls, slabs, beams

2.4.14.2 Staff Evaluation

The staff reviewed LRA Section 2.4.14 using the evaluation methodology of SER Section 2.4. The staff conducted its review in accordance with the guidance of SRP-LR Section 2.4.

In conducting its review, the staff evaluated the structural component functions described in the LRA and UFSAR to verify that the applicant had not omitted from the scope of license renewal any components with intended functions under 10 CFR 54.4(a). The staff then reviewed those components that the applicant had identified as within the scope of license renewal to verify that it had not omitted any passive and long-lived components subject to an AMR in accordance with the requirements of 10 CFR 54.21(a)(1).

2.4.14.3 Conclusion

The staff reviewed the LRA and related structural components to determine whether any SSCs that should be within the scope of license renewal had not been identified by the applicant. No omissions were identified. In addition, the staff determined whether any components subject to an AMR had not been identified by the applicant. No omissions were identified. The staff concludes that there is reasonable assurance that the applicant has adequately identified the office building components within the scope of license renewal, as required by 10 CFR 54.4(a), and those subject to an AMR, as required by 10 CFR 54.21(a)(1).

2.4.15 Oyster Creek Substation

2.4.15.1 Summary of Technical Information in the Application

In LRA Section 2.4.15, the applicant described the OCGS substation located west of the reactor building adjacent to the intake and discharge canals. The purpose of the substation is to provide structural support, shelter, and protection to nonsafety-related electrical components and commodities. The substation consists of a reinforced concrete slab on grade, the breaker switch control room, transmission towers, and the foundation for OCGS output power to the grid and for incoming offsite power system components. The breaker switch control room is a commercial grade steel enclosure with metal siding and metal deck supported on the substation concrete slab. The substation is classified as nonsafety-related, seismic Class II.

The OCGS substation performs functions that support SBO.

The intended functions within the scope of license renewal include:

- provides enclosure, shelter, or protection for in-scope equipment (including shielding)
- provides structural support or structural integrity to preclude nonsafety-related component interactions that could prevent satisfactory accomplishment of a safety-related function

In LRA Table 2.4.15, the applicant identified the following OCGS substation component types within the scope of license renewal and subject to an AMR:

- conduits
- door
- equipment foundation
- metal deck
- metal siding
- reinforced concrete foundation
- seals
- structural bolts
- structural steel
- transmission towers

2.4.15.2 Staff Evaluation

The staff reviewed LRA Section 2.4.15 using the evaluation methodology of SER Section 2.4. The staff conducted its review in accordance with the guidance of SRP-LR Section 2.4.

In conducting its review, the staff evaluated the structural component functions described in the LRA and UFSAR to verify that the applicant had not omitted from the scope of license renewal any components with intended functions under 10 CFR 54.4(a). The staff then reviewed those components that the applicant had identified as within the scope of license renewal to verify that it had not omitted any passive and long-lived components subject to an AMR in accordance with the requirements of 10 CFR 54.21(a)(1).

2.4.15.3 Conclusion

The staff reviewed the LRA and related structural components to determine whether any SSCs that should be within the scope of license renewal had not been identified by the applicant. No omissions were identified. In addition, the staff determined whether any components subject to an AMR had not been identified by the applicant. No omissions were identified. The staff concludes that there is reasonable assurance that the applicant has adequately identified the OCGS substation components within the scope of license renewal, as required by 10 CFR 54.4(a), and those subject to an AMR, as required by 10 CFR 54.21(a)(1).

2.4.16 Turbine Building

2.4.16.1 Summary of Technical Information in the Application

In LRA Section 2.4.16, the applicant described the turbine building as a reinforced concrete and steel structure directly west of the reactor building and adjacent to the office building. The purpose of the building is to provide structural support, shelter, and protection for safety-related and nonsafety-related SSCs housed within. The building contains the plant control room, two cable spreading rooms, the 4160V switchgear room, the "C" battery room, and a mechanical equipment room (HVAC) for the control room. The control room, the two cable spreading rooms, and the mechanical equipment room on the northeast corner of the building are enclosed in reinforced concrete walls and slabs to protect safety-related components and control room personnel from extreme environmental conditions and DBEs. The rest of the building encloses the steam and power conversion system, the TBCCW system, reactor protection system components, turbine building ventilation, the hydrogen injection system, and supporting systems. Major components within the building include turbine generators, main condensers, moisture separators, reheaters, reactor feedwater pumps, main steam control and stop valves, condensate pumps, TBCCW heat exchangers, and their piping. Highly radioactive components are enclosed within heavy concrete walls with labyrinthine entrances for shielding purposes. Equipment in the building is serviced by two cranes, the turbine building overhead bridge crane and the heater bay overhead bridge crane. The building foundation is a reinforced concrete mat founded on dense Cohansey sand 31 feet below grade level. Reinforced concrete walls extend from the top of the base mat level to the turbine generator operating floor 23 feet above grade level. Steel framework and insulated metal siding and built-up roofing enclose the turbine generator operating floor.

The turbine building contains safety-related components relied upon to remain functional during and following DBEs. The failure of nonsafety-related SSCs in the turbine building potentially could prevent the satisfactory accomplishment of a safety-related function. In addition, the turbine building performs functions that support fire protection, ATWS, SBO, and EQ.

The intended functions within the scope of license renewal include:

- provides enclosure, shelter, or protection for in-scope equipment (including shielding)
- provides flood protection barrier (internal and external flood event)
- provides HELB shielding

2-159

- provides missile barrier (internal or external)
- provides shielding against radiation
- provides structural support or structural integrity to preclude nonsafety-related component interactions that could prevent satisfactory accomplishment of a safety-related function

In LRA Table 2.4.16, the applicant identified the following turbine building component types within the scope of license renewal and subject to an AMR:

- bird screen
- cable tray
- concrete embedments
- conduits
- equipment foundation
- hatch plugs
- masonry block walls
- metal deck
- metal siding
- miscellaneous steel (catwalks, handrails, ladders, platforms, grating, and associated supports)
- panels and enclosures
- penetration seals
- reinforced concrete foundation
- reinforced concrete walls (above and below grade)
- reinforced concrete walls, slabs, beams
- roofing
- seals
- structural bolts
- structural steel: beams, columns, girders, plate

2.4.16.2 Staff Evaluation

The staff reviewed LRA Section 2.4.16 using the evaluation methodology of SER Section 2.4. The staff conducted its review in accordance with the guidance of SRP-LR Section 2.4.

In conducting its review, the staff evaluated the structural component functions described in the LRA and UFSAR to verify that the applicant had not omitted from the scope of license renewal any components with intended functions under 10 CFR 54.4(a). The staff then reviewed those components that the applicant had identified as within the scope of license renewal to verify that it had not omitted any passive and long-lived components subject to an AMR in accordance with the requirements of 10 CFR 54.21(a)(1).

2.4.16.3 Conclusion

The staff reviewed the LRA and related structural components to determine whether any SSCs that should be within the scope of license renewal had not been identified by the applicant. No omissions were identified. In addition, the staff determined whether any components subject to an AMR had not been identified by the applicant. No omissions were identified. The staff concludes that there is reasonable assurance that the applicant has adequately identified the turbine building components within the scope of license renewal, as required by 10 CFR 54.4(a), and those subject to an AMR, as required by 10 CFR 54.21(a)(1).

2.4.17 Ventilation Stack

2.4.17.1 Summary of Technical Information in the Application

In LRA Section 2.4.17, the applicant described the ventilation stack as a 394-foot high, tapered, reinforced concrete structure southeast of the reactor building and adjacent to the SGTS and the heating boiler house. The purpose of the ventilation stack is to provide an elevated discharge point for gaseous effluents collected from the SGTS, RBVS, radwaste area heating and ventilation system, main condenser air extraction system (includes turbine steam seal effluents), augmented offgas system, and turbine building ventilation system. In addition, the stack in conjunction with the hardened vent system provides a secondary pressure vent path for primary containment if the torus vent path is unavailable. Effluents through the ventilation stack are monitored to ensure that the 10 CFR Part 20 limits, which apply to releases during normal operation, and the 10 CFR Part 100 limits, which apply to accidental releases, are not exceeded. The stack also provides structural support to the piping, tubing, and air ducts penetrating it and to components inside it, including valves, absolute filter, and radiation monitors. Its base is a 7-foot thick reinforced concrete slab founded on very dense sand and buried 26 feet below grade. Internally, the structure is divided into three levels formed by the base slab, an intermediate slab at ground level, and an upper slab located 11' 6" above ground level. Access into the stack is from the old heating boiler house and from the exhaust tunnel. The stack is classified as seismic Class I and relied upon to elevate gaseous effluents during normal plant operation and during accident conditions.

The ventilation stack contains safety-related components relied upon to remain functional during and following DBEs. The failure of nonsafety-related SSCs in the ventilation stack potentially could prevent the satisfactory accomplishment of a safety-related function.

The intended functions within the scope of license renewal include:

- provides path for release of filtered and unfiltered gaseous discharge
- maintains mechanical and structural integrity to prevent spatial interactions that could cause failure of safety-related SSCs (includes the required structural support when the nonsafety-related leakage boundary piping is also attached to safety-related piping)
- provides structural support or structural integrity to preclude nonsafety-related component interactions that could prevent satisfactory accomplishment of a safety-related function

In LRA Table 2.4.17, the applicant identified the following ventilation stack component types within the scope of license renewal and subject to an AMR:

- concrete embedments
- hatch cover
- miscellaneous steel (catwalks, handrails, ladders, platforms, grating, and associated supports)

- penetration seals
- penetration sleeve, cap plates, capped auxiliary boiler exhaust pipe
- reinforced concrete foundation
- reinforced concrete slabs
- reinforced concrete stack (above and below grade)
- structural bolts

2.4.17.2 Staff Evaluation

The staff reviewed LRA Section 2.4.17 using the evaluation methodology of SER Section 2.4. The staff conducted its review in accordance with the guidance of SRP-LR Section 2.4.

In conducting its review, the staff evaluated the structural component functions described in the LRA and UFSAR to verify that the applicant had not omitted from the scope of license renewal any components with intended functions under 10 CFR 54.4(a). The staff then reviewed those components that the applicant had identified as within the scope of license renewal to verify that it had not omitted any passive and long-lived components subject to an AMR in accordance with the requirements of 10 CFR 54.21(a)(1).

2.4.17.3 Conclusion

The staff reviewed the LRA and related structural components to determine whether any SSCs that should be within the scope of license renewal had not been identified by the applicant. No omissions were identified. In addition, the staff determined whether any components subject to an AMR had not been identified by the applicant. No omissions were identified. The staff concludes that there is reasonable assurance that the applicant has adequately identified the ventilation stack components within the scope of license renewal, as required by 10 CFR 54.4(a), and those subject to an AMR, as required by 10 CFR 54.21(a)(1).

2.4.18 Component Supports Commodity Group

2.4.18.1 Summary of Technical Information in the Application

In LRA Section 2.4.18, the applicant described the component supports commodity group consisting of structural elements and specialty components designed to transfer the load applied from an SSC to building structural elements or directly to building foundations. Supports include seismic anchors or restraints, frames, constant and variable spring hangers, rod hangers, sway struts, guides, stops, design clearances, straps, clamps, and clevis pins. Specialty components include snubbers, sliding surfaces, and vibration isolators. Sliding surfaces, when incorporated into the support design, permit release of lateral forces but are relied upon to carry vertical load. Specialty supports like snubbers only resist seismic forces. Vibration isolators are incorporated in the design of some vibrating equipment to minimize the impact of vibration. Other support types like guides and position stops allow displacement in a specified direction or preclude unacceptable movements and interactions.

The commodity group is comprised of the following supports:

- supports for ASME Class 1, 2 and 3 piping and components including reactor vessel stabilizer, reactor vessel skirt support, and CRD housing supports
- supports for ASME Class MC components including suppression chamber seismic restraints, suppression chamber support saddles and columns, and vent system supports
- supports for cable trays, conduit, HVAC ducts, tube track, and instrument tubing
- supports for non-ASME piping and components including EDG supports
- supports for racks, panels, and enclosures
- supports for spray shields and masonry walls

The component supports commodity group contains safety-related components relied upon to remain functional during and following DBEs. The failure of nonsafety-related SSCs in the component supports commodity group potentially could prevent the satisfactory accomplishment of a safety-related function. In addition, the component supports commodity group performs functions that support fire protection, ATWS, SBO, and EQ.

The intended functions within the scope of license renewal include:

- provides structural support or structural integrity to preclude nonsafety-related component interactions that could prevent satisfactory accomplishment of a safety-related function
- provides flexible support for HVAC fan units

In LRA Table 2.4.18, the applicant identified the following component supports commodity group component types within the scope of license renewal and subject to an AMR:

- building concrete at locations of expansion and grouted anchors, grouted pads for support base plates
- supports for ASME Class 1 piping and components (constant and variable load spring hangers, guides, stops, sliding surfaces, design clearances)
- supports for ASME Class 1 piping and components (support members, welds, bolted connections, support anchorage to building structure)
- supports for ASME Class 2 and 3 piping and components (constant and variable load spring hangers, guides, stops, sliding surfaces, design clearances)
- supports for ASME Class 2 and 3 piping and components (support members, welds, bolted connections, support anchorage to building structure)
- supports for ASME Class MC components (guides, stops, sliding surfaces, design clearances)
- supports for ASME Class MC components (support members, welds, bolted connections, support anchorage to building structure)
- supports for cable trays (support members, welds, bolted connections, support anchorage to building structure)
- supports for conduits (support members, welds, bolted connections, support anchorage to building structure)

2-163

- supports for HVAC components (vibration isolation elements)
- supports for HVAC components and other miscellaneous mechanical equipment (support members, welds, bolted connections, support anchorage to building structure)
- supports for HVAC ducts (support members, welds, bolted connections, support anchorage to building structure)
- supports for masonry walls (support members, welds, bolted connections, support anchorage to building structure)
- supports for non-ASME piping and components (support members, welds, bolted connections, support anchorage to building structure)
- supports for panels and enclosures, racks (support members, welds, bolted connections, support anchorage to building structure)
- supports for platforms, pipe whip restraints, jet impingement and spray shields, and other miscellaneous structures (support members, welds, bolted connections, support anchorage to building structure)
- supports for tube track and instrument tubing (support members, welds, bolted connections, support anchorage to building structure)

2.4.18.2 Staff Evaluation

The staff reviewed LRA Section 2.4.18 using the evaluation methodology of SER Section 2.4. The staff conducted its review in accordance with the guidance of SRP-LR Section 2.4.

In conducting its review, the staff evaluated the structural component functions described in the LRA and UFSAR to verify that the applicant had not omitted from the scope of license renewal any components with intended functions under 10 CFR 54.4(a). The staff then reviewed those components that the applicant had identified as within the scope of license renewal to verify that it had not omitted any passive and long-lived components subject to an AMR in accordance with the requirements of 10 CFR 54.21(a)(1).

2.4.18.3 Conclusion

The staff reviewed the LRA and related structural components to determine whether any SSCs that should be within the scope of license renewal had not been identified by the applicant. No omissions were identified. In addition, the staff determined whether any components subject to an AMR had not been identified by the applicant. No omissions were identified. The staff concludes that there is reasonable assurance that the applicant has adequately identified the component supports commodity group components within the scope of license renewal, as required by 10 CFR 54.4(a), and those subject to an AMR, as required by 10 CFR 54.21(a)(1).

2.4.19 Piping and Component Insulation Commodity Group

2.4.19.1 Summary of Technical Information in the Application

In LRA Section 2.4.19, the applicant described the piping and component insulation commodity group comprised of pre-fabricated blankets, modules, or panels engineered as integrated assemblies to fit the surface to be insulated and to fit easily against the piping and components.

The insulation includes originally installed and replacement metallic and nonmetallic materials. The purpose of insulation is to improve thermal efficiency, minimize heat loads on the HVAC systems, provide protection for personnel, or prevent sweating of cold piping and components. Metallic insulation consists of stainless steel mirror insulation. Nonmetallic insulation consists of calcium silicate, asbestos, and light-density, semi-rigid fibrous glass quilted between two layers of glass scrim and encapsulated in a fiberglass cloth forming a composite blanket or of pre-molded fiberglass modules and panels encased in fiberglass jackets. Anti-sweat insulation consists of closed cell, foamed plastic (inside primary containment drywell) and fiberglass dual-temperature or glass wool blanketing (outside primary containment drywell). Metal protective jackets are made from rolled aluminum or stainless steel. The insulation is a nonsafety-related commodity.

The failure of nonsafety-related SSCs in the piping and component insulation commodity group potentially could prevent the satisfactory accomplishment of a safety-related function.

The intended functions within the scope of license renewal include:

- provides physical support of thermal insulation and prevents moisture absorption
- provides heat loss control to preclude overheating of nearby safety-related SSCs

In LRA Table 2.4.19, the applicant identified the following piping and component insulation commodity group component types within the scope of license renewal and subject to an AMR:

- insulation
- insulation jacketing

2.4.19.2 Staff Evaluation

The staff reviewed LRA Section 2.4.19 using the evaluation methodology of SER Section 2.4. The staff conducted its review in accordance with the guidance of SRP-LR Section 2.4.

In conducting its review, the staff evaluated the structural component functions described in the LRA and UFSAR to verify that the applicant had not omitted from the scope of license renewal any components with intended functions under 10 CFR 54.4(a). The staff then reviewed those components that the applicant had identified as within the scope of license renewal to verify that it had not omitted any passive and long-lived components subject to an AMR in accordance with the requirements of 10 CFR 54.21(a)(1).

2.4.19.3 Conclusion

The staff reviewed the LRA and related structural components to determine whether any SSCs that should be within the scope of license renewal had not been identified by the applicant. No omissions were identified. In addition, the staff determined whether any components subject to an AMR had not been identified by the applicant. No omissions were identified. The staff concludes that there is reasonable assurance that the applicant has adequately identified the piping and component insulation commodity group components within the scope of license renewal, as required by 10 CFR 54.4(a), and those subject to an AMR, as required by 10 CFR 54.21(a)(1).

2.5 Scoping and Screening Results: Electrical Components

This section documents the staff's review of the applicant's scoping and screening results for electrical and instrumentation and control (I&C) systems. Specifically, this section discusses the electrical and I&C systems and the electrical commodity groups.

In accordance with the requirements of 10 CFR 54.21(a)(1), the applicant must identify and list passive, long-lived SCs within the scope of license renewal and subject to an AMR. To verify that the applicant properly implemented its methodology, the staff focused its review on the implementation results. This focus allowed the staff to confirm that there were no omissions of electrical and I&C system components meeting the scoping criteria and subject to an AMR.

Staff Evaluation Methodology. The staff's evaluation of the information provided in the LRA was the same for all electrical and I&C systems. The objective was to determine whether the components and supporting structures for a specific system or commodity group, that appeared to meet the scoping criteria specified in the Rule, had been identified by the applicant as within the scope of license renewal, in accordance with 10 CFR 54.4. Similarly, the staff evaluated the applicant's screening results to verify that all long-lived, passive components were subject to an AMR in accordance with 10 CFR 54.21(a)(1).

Scoping. For its evaluation, the staff reviewed the applicable LRA sections and associated component drawings, focusing its review on components that had not been identified as within the scope of license renewal. The staff reviewed relevant licensing basis documents, including the UFSAR, for each system and commodity group to determine whether the applicant had omitted components with intended functions under 10 CFR 54.4(a) from the scope of license renewal. The staff also reviewed the licensing basis documents to determine whether all intended functions under 10 CFR 54.4(a) were specified in the LRA. If omissions were identified, the staff requested additional information to resolve them.

Screening. After completing its review of the scoping results, the staff evaluated the applicant's screening results. For those SCs with intended functions, the staff sought to determine whether (1) the functions are performed with moving parts or a change in configuration or properties, or (2) they are subject to replacement based on a qualified life or specified time period, as described in 10 CFR 54.21(a)(1). For those that did not meet either of these criteria, the staff sought to confirm that these SCs were subject to an AMR, as required by 10 CFR 54.21(a)(1). If discrepancies were identified, the staff requested additional information to resolve them.

2.5.1 Summary of Technical Information in the Application

2.5.1.1 Electrical Systems

In LRA Section 2.5.1, the applicant described the electrical and I&C systems. The electrical systems include the following:

- 120/208V non-essential distribution system
- 120V AC vital power system
- 125V station DC system
- 24/48V instrument power DC system
- 4160V AC system
- 480/208/120V utility (JCP&L) non-vital power

- 480V AC system
- alternate rod injection system
- grounding and lightning protection system
- intermediate range monitoring system
- lighting system
- local power range monitoring system and average power range monitoring system
- offsite power system
- post-accident monitoring system
- radio communications system
- reactor overfill protection system
- reactor protection system
- remote shutdown system
- SBO system

The electrical systems contain safety-related components relied upon to remain functional during and following DBEs. The failure of nonsafety-related SSCs in the electrical systems potentially could prevent the satisfactory accomplishment of a safety-related function. In addition, the electrical systems perform functions that support fire protection, ATWS, SBO, and EQ.

120/208V Non-Essential Distribution System. The 120/208V non-essential electrical distribution system receives power from 460V motor control centers and 460V distribution panels through dry-type transformers. The system is designed to provide nonessential power to the various nonsafety-related and auxiliary plant loads. Additional detail of the system is in UFSAR Section 8.3.1.1.3.

This system is within the scope of license renewal because it (a) resists nonsafety-related SSC failures that could prevent satisfactory accomplishment of a safety-related function (this system provides electrical power to a control room ventilation fan) and (b) is relied upon in safety analyses or plant evaluations to perform a function for compliance with fire protection and SBO regulations.

120V AC Vital Power System. The 120V AC vital power system is a Class 1E safety-related electrical distribution system that supplies 120V AC power to various loads essential for operation, protection, and safe shutdown of the plant. The system design incorporates redundant power sources and automatic bus transfer switches so that critical loads remain energized at all times. Additional detail of the system is in UFSAR Section 8.3.1.1.4.

This system is within the scope of license renewal because it (a) provides motive power to safety-related components and (b) is relied upon in safety analyses or plant evaluations to perform a function for compliance with fire protection, EQ, and SBO regulations.

125V Station DC System. Three complete 125V DC distribution systems make up the station DC power system at OCGS. Two of these, designated as DC Distribution Systems A and B, are the originally installed systems. The third system, designated as DC Distribution System C, was designed and installed as a modification.

The function of the station DC system is to provide a continuous source of 125V DC power. Safety loads are supplied from DC Distribution Systems B and C with DC Distribution System B supplying Division B safety-related loads and DC Distribution System C supplying Division A safety-related loads. DC Distribution System A supplies nonsafety loads. Additional detail of the

2-167

system is in UFSAR Section 8.3.2.1.

This system is within the scope of license renewal because it (a) provides motive power to safety-related components and (b) is relied upon in safety analyses or plant evaluations to perform a function for compliance with fire protection, EQ, and SBO regulations.

24/48V Instrumentation Power DC System. The 24/48V DC power electrical distribution system is designed to supply power to the reactor nuclear instrumentation and radiation monitoring systems. Additional detail of the system is in UFSAR Section 8.3.2.2.

This system is within the scope of license renewal because provides motive power to safety-related components.

4160V System. The 4160V electrical distribution system is designed to provide continuous electrical power necessary for plant operation, startup, and shutdown. The 4160V switchgear is comprised of four separate bus sections or lineups of switchgear. The four bus sections are identified as Bus Sections 1A, 1B, 1C, and 1D with Bus Sections 1C and 1D being the essential or emergency switchgear lineups.

The 4160V AC system also can be powered from the FRCT, which is the OCGS alternate AC (AAC) power source during an SBO event. The AAC source utilizes a connection independent from the normal connection to the regional transmission grid. The routing is through a dedicated underground ductbank to the load break switches and SBO transformer located on site and then through a cable trench to the switchgear breaker connection to the 4160V AC Bus 1B. Additional detail of the system is in UFSAR Section 8.3.1.1.1.

This system is within the scope of license renewal because it (a) provides motive power to safety-related components and (b) is relied upon in safety analyses or plant evaluations to perform a function for compliance with fire protection and SBO regulations.

480/208/120V Utility (JCP&L) Non-Vital Power System. The 480/208/120V utility (JCP&L) nonvital power electrical distribution system is designed to provide nonessential electrical power necessary for balance of plant equipment located throughout the site. Additional detail of the system is in UFSAR Section 8.2.1.2.

This system is within the scope of license renewal because it is relied upon in safety analyses or plant evaluations to perform a function for compliance with fire protection regulations.

480V AC System. The 480V AC electrical distribution system is designed to provide continuous electrical power necessary for plant operation, startup, and shutdown. Additional detail of the system is in UFSAR Section 8.3.1.1.2.

This system is within the scope of license renewal because it (a) provides motive power to safety-related components and (b) is relied upon in safety analyses or plant evaluations to perform a function for compliance with fire protection, EQ, and SBO regulations.

Alternate Rod Injection System. The alternate rod injection electrical system provides a method diverse from the reactor protection system (RPS) for depressurizing the instrument (control) air system scram air header in the unlikely event the RPS does not cause a reactor scram in response to an operational transient. Additional detail of the system is in UFSAR Section 3.9.4.4.

This system is within the scope of license renewal because it is relied upon in safety analyses or plant evaluations to perform a function for compliance with ATWS regulations.

Grounding and Lightning Protection System. The plant grounding and lightning protection electrical system is designed to provide a low-impedance path to ground for fault currents and lightning strokes.

This system is within the scope of license renewal because it is relied upon in safety analyses or plant evaluations to perform a function for compliance with fire protection regulations.

Intermediate Range Monitoring System. The intermediate range monitoring electrical instrumentation and logic system is designed to monitor the neutron flux and power in the reactor core and to provide automatic core protection. The intermediate range monitoring system provides the operator with power level indication and generates annunciator alarms, rod blocks, and scram signals for nuclear instrumentation degraded operation and downscale or upscale conditions. Additional detail of the system is in UFSAR Section 7.5.1.8.4.

This system is within the scope of license renewal because it senses process conditions and generates signals for a reactor trip or an ESF actuation.

Lighting System. The lighting system is comprised of the normal lighting and convenience system (outdoor area lighting, general plant lighting, office building lighting), emergency lighting, and security lighting. Additional detail of the system is in UFSAR Section 9.5.3.

This system is within the scope of license renewal because it is relied upon in safety analyses or plant evaluations to perform a function for compliance with fire protection and SBO regulations.

Local Power Range Monitoring System and Average Power Range Monitoring System. The local power range and average power range monitoring electrical instrumentation and logic systems are designed to monitor the neutron flux and power in the reactor core and to provide automatic core protection. Additional detail of the system is in UFSAR Sections 7.5.1.8.6 and 7.5.1.8.7.

This system is within the scope of license renewal because it senses process conditions and generates signals for a reactor trip or an ESF actuation.

Offsite Power System. The offsite power electrical distribution system is designed to connect OCGS to the offsite electrical transmission system. The purpose of the offsite power system is to connect to the output of the generator and to provide redundant sources of power to the plant when the main generator is offline. It accomplishes this purpose with a 230 kV substation and a connected 34.5 kV substation in a switchyard adjacent to the plant. Additional detail of the system is in UFSAR Section 8.2.

This system is within the scope of license renewal because it is relied upon in safety analyses or plant evaluations to perform a function for compliance with fire protection and SBO regulations.

Post-Accident Monitoring System. The purpose of the post-accident electrical monitoring system is to display and record plant parameters of drywell radiation and pressure levels, torus level, and temperature and safety/relief valve flow detection during and following a LOCA. The system is comprised of containment high-range radiation monitors, safety valve and relief valve accident monitoring instrumentation, suppression pool temperature and water level monitors, and containment pressure indicators. Additional detail of the system is in UFSAR Sections 1.9, 12.3.4.1.5, 5.2.2.4.2.2, 7.6.1.4, and 11.5.2.13.

This system is within the scope of license renewal because it (a) senses process conditions and generates signals for a reactor trip or an ESF actuation and (b) is relied upon in safety analyses or plant evaluations to perform a function for compliance with fire protection and EQ regulations.

Radio Communications System. The radio communications electrical system is designed to provide two-way voice communication between personnel operating safe shutdown equipment during a fire emergency and SBO. The radio communications system is comprised of primary and installed spare base station transmitter-repeaters in the upper cable spreading room, portable radio units with batteries and chargers in the control room, and antennae with associated cabling at selected locations in the reactor building and turbine building. Electrical power for the primary base station transmitter and repeater is supplied from the 120V AC vital power system.

This system is within the scope of license renewal because it is relied upon in safety analyses or plant evaluations to perform a function for compliance with fire protection and SBO regulations (See SER Section 3.7 for additional information on the Radio Communications System as it related to the Meteorological Tower.).

Reactor Overfill Protection System. The reactor overfill protection electrical instrumentation and logic system minimizes the potential for overfilling the reactor to the elevation of the main steam lines. Additional detail of the system is in UFSAR Section 7.7.1.6.

This system is within the scope of license renewal because failure of its components could adversely affect the safety-related RPS.

Reactor Protection System. The RPS is an electrical logic system designed to furnish signals to trip the reactor and to initiate certain ESF systems. Additional detail of the system is in UFSAR Sections 7.2 and 7.3.

This system is within the scope of license renewal because it (a) senses process conditions and generates signals for a reactor trip or an ESF actuation and (b) is relied upon in safety analyses or plant evaluations to perform a function for compliance with fire protection regulations.

Remote Shutdown System. The remote shutdown system enables operators to achieve and maintain hot and cold shutdown whenever it is necessary to evacuate the control room. The remote shutdown system is comprised of a remote shutdown panel and several local shutdown panels outside the control room. Additional detail of the system is in UFSAR Sections 9.5.1 and 3.1.15.

This system is within the scope of license renewal because it (a) monitors conditions and controls plant equipment to achieve and maintain safe shutdown and senses process conditions and generates signals for a reactor trip or an ESF actuation and (b) is relied upon in safety analyses or plant evaluations to perform a function for compliance with fire protection, EQ, and SBO regulations.

Station Blackout System. The SBO electrical supply system provides AAC power for the regulated event of loss of all AC power. The source of electrical power to the SBO system is the

FRCT station, an electrical power plant owned, operated, and maintained by FirstEnergy and designed for peak loading to the grid. Additional detail of the system is in UFSAR Sections 8.3.4 and 15.9.

This system is within the scope of license renewal because it is relied upon in safety analyses or plant evaluations to perform a function for compliance with SBO regulations.

2.5.1.2 Electrical Commodity Groups

In LRA Section 2.5.2.5, the applicant described the electrical commodity groups subject to an AMR. The screening process for electrical components used plant documentation to identify the electrical component types within the electrical, mechanical, and civil or structural systems based on plant design documentation, drawings, the CRL, and interface with the parallel mechanical and civil screening efforts. These component types were grouped into a smaller set of electrical commodity groups identified from a review of NEI 95-10 Appendix B, the GALL Report, and information from previous LRAs.

The intended functions within the scope of license renewal include:

- provides electrical continuity
- provides insulation and support for an electric conductor
- provides pressure-retaining boundary; fission product barrier; containment isolation

In LRA Table 2.5.2, the applicant identified the following electrical commodity group component types within the scope of license renewal and subject to an AMR:

- cable connections (metallic parts)
- electrical penetrations
- fuse holders
- high-voltage insulators
- insulated cables and connections
- insulated cables and connections in instrumentation circuits
- insulated inaccessible medium-voltage cables
- transmission conductors and connections
- uninsulated ground conductors
- wooden utility poles

The phase bus in the main generator and auxiliaries system and the switchyard bus were not included within the AMRs because they perform no license renewal intended function. The phase bus is further discussed in SER Section 2.5.3.

The commodity groups were screened by 10 CFR 54.21(a)(1)(ii) criteria that allow the exclusion of component commodity groups subject to replacement based on a qualified life or specified time period. The only electrical components excluded by the 10 CFR 54.21(a)(1)(ii) criteria are included in the Environmental Qualification Program because they are replaced prior to the expiration of their defined qualified lives. No electrical components within the Environmental Qualification Program are subject to an AMR by 10 CFR 54.21(a)(1)(ii) screening criteria. Therefore, the electrical components in the Environmental Qualification Program were screened out.

The remaining commodity groups, some or all of which are not in the Environmental Qualification

Program, are within the scope of license renewal and require an AMR. In the LRA, the following commodity groups are discussed:

(1) Insulated Cables and Connections - The insulated cables and connections commodity group was broken down for an AMR of insulation into subcategories based on their treatment in the GALL Report:

- insulated cables and connections
- insulated cables and connections in instrumentation circuits
- insulated inaccessible medium-voltage cables

The types of insulated connections included in this review are splices, connectors, and terminal blocks. Fuse holders were reviewed separately.

(2) Electrical Penetrations - The electrical portions of those electrical penetrations not included in the Environmental Qualification Program meet the 10 CFR 54.21(a)(1)(ii) screening criterion and are subject to an AMR. The electrical insulation within the penetration assembly and the epoxy potting compound that provides the sealing function were reviewed. Insulated cable pigtails are considered part of the insulated cables and connectors commodity group. Metallic portions of the electrical penetrations are considered part of the primary containment structure.

(3) High Voltage Insulators - High-voltage insulators are on the circuits supplying power from the switchyard to plant buses during recovery from an SBO or fire protection event. The high-voltage insulators meet the 10 CFR 54.21(a)(1)(ii) screening criterion and are subject to an AMR.

(4) Transmission Conductors and Connections - Transmission conductors that provide a portion of the circuits supplying power from the switchyard to plant buses during recovery from an SBO or fire protection event meet the 10 CFR 54.21(a)(1)(ii) screening criterion and are subject to an aging management review.

(5) Fuse Holders - Both the metallic and nonmetallic portions of fuse holders not included in the Environmental Qualification Program meet the 10 CFR 54.21(a)(1)(ii) screening criterion and are subject to an AMR.

(6) Wooden Utility Poles - Wooden utility poles did not fit within an existing electrical commodity group; therefore, a separate commodity group was created. Utility poles provide structural support for transmission conductors, high-voltage insulators, and other active electrical components supplying power from the switchyard to plant buses during recovery from an SBO or fire protection event. The wooden utility poles meet the 10 CFR 54.21(a)(1)(ii) screening criterion and are subject to an AMR.

(7) Cable Connections (Metallic Parts) - The cable connections commodity group includes the metallic portions of cable connections not included in the Environmental Qualification Program. The metallic connections evaluated include splices, threaded connectors, compression type termination lugs, and terminal blocks.

(8) Uninsulated Ground Conductors - The uninsulated ground conductors commodity group is comprised of grounding cable and connectors.

The components which support or interface with electrical components (e.g., cable trays, conduits, instrument racks, panels, and enclosures) are assessed as part of the structures in which they are located, as discussed in LRA Section 2.4

2.5.2 Staff Evaluation

The staff reviewed LRA Section 2.5 and the UFSAR using the evaluation methodology of SER Section 2.5. The staff conducted its review in accordance with the guidance of SRP-LR Section 2.5.

In conducting its review, the staff evaluated the system functions described in the LRA and UFSAR to verify that the applicant had not omitted from the scope of license renewal any components with intended functions under 10 CFR 54.4(a). The staff then reviewed those components that the applicant had identified as within the scope of license renewal to verify that the applicant had not omitted any passive and long-lived components subject to an AMR in accordance with 10 CFR 54.21(a)(1).

The staff's review of LRA Section 2.5 identified areas in which additional information was necessary to complete the review of the applicant's scoping and screening results. The applicant responded to the staff's RAIs as discussed below.

In RAI 2.5.1.19-1 dated September 28 2005, the staff stated that the combustion turbine power plant was determined to be within the scope of license renewal. The staff requested that the applicant evaluate the long-lived passive components of the combustion turbine power plant and any AMPs and AMRs related to those components in the same format and depth as used in the diesel generator section of the LRA.

In its response dated October 12, 2005, the applicant stated:

> AmerGen has taken a more detailed approach to scoping, screening, aging management reviews and aging management programs, for long-lived passive components, than was previously presented in the Oyster Creek License Renewal Application submittal for the Oyster Creek Station Blackout System, Combustion Turbine Power Plant.

In addition, the applicant revised Commitment Nos. 31 and 36. Furthermore, Commitment No. 43, "Periodic Monitoring of Combustion Turbine - Electrical," was completely modified as follows:

> A new plant specific program, 'Periodic Monitoring of Combustion Turbine Power Plant - Electrical' is credited. The program will be used in conjunction with the existing 'Structures Monitoring Program' and the new 'Inaccessible Medium Voltage Cables Not Subject to 10 CFR 50.59 Environmental Qualification Requirements Program', to manage the aging effects for the electrical commodities that support Forked River Combustion Turbine (FRCT) operation. The Program consists of visual inspections of accessible electrical cables and connections exposed in enclosures, pits, manholes and pipe trench; visual inspection for water collection in manholes, pits, and trenches, located on the FRCT site, for inaccessible medium voltage cables; and visual inspection of

accessible phase bus and connections and phase bus insulators/supports. The new program will be performed on a 2-year interval for manhole, pit and trench inspections, on a 5-year interval for phase bus inspections, and on a 10-year interval for cable and connection inspections.

In Appendix B of this letter, the applicant described the scoping system in more detail, correlating to LRA Section 2.5.1.19, "Station Blackout," for scoping and screening results. Sixteen subsystem descriptions (e.g., fuel oil system, combustion turbine inlet and exhaust system, cooling water system), combustion turbine structure and electrical commodity descriptions, and associated system boundary details have been added to the scoping information. The applicant stated that the expanded information is consistent with such other LRA system information as the EDGs.

The applicant identified and described the following SBO system electrical commodity groups subject to AMR in Section 2.5.2A.5 of its letter:

•	cable connections (metallic parts)	(Section 2.5.2A.5.1)
•	high-voltage insulators	(Section 2.5.2A.5.2)
•	insulated cables and connections	(Section 2.5.2A.5.3)
•	phase bus	(Section 2.5.2A.5.4)
•	transmission conductors and connections	(Section 2.5.2A.5.5)
•	uninsulated ground conductors	(Section 2.5.2A.5.6)

The staff reviewed the applicant's response following the guidance of SRP-LR, Section 2.5. The staff agreed that the electrical commodities groups in the SBO recovery path consisting of passive long-lived components subject to AMR are in accordance with 10 CFR 54.21(a)(1).

In RAI 2.5.2-1 dated March 20, 2006, the staff noted that LRA Section 2.5.2.5 describes electrical commodity groups subject to an AMR. The staff requested that the applicant confirm that, in addition to power circuits in the electrical systems, the control circuits also had been considered in the scoping and screening review and included in the electrical commodity groups subject to AMR.

In its response dated April 18, 2006, the applicant clarified that both power and control circuits had been considered in the scoping, screening, AMR, and AMP processes for the electrical commodity groups.

In RAI 2.5.2.3-1 dated March 20, 2006, the staff noted that in LRA Section 2.5.2.3 the first bullet states: "Phase Bus exist only in the Main Generator and Auxiliaries System. The system has no electrical intended functions and is within the scope for 10 CFR 54.4(a)(2) systems interaction only. Because the phase bus contains no fluid, it has no license renewal intended functions."

The staff requested that the applicant address the following as to that statement:

• Provide a cross-reference to the phase bus in the SBO path.

• Confirm whether the phase bus (in the main generator and auxiliaries system) provides interactions for 10 CFR 54.4(a)(2) systems. If yes, list the 10 CFR 54.4(a)(2) systems. If 10 CFR 54.4(a)(2) applies to this phase bus, explain why it is not included as an electrical commodity group subject to an AMR.

- Explain the statement: "Because the phase bus contains no fluid, it has no license renewal intended functions."

In its response dated April 18, 2006, the applicant stated that as part of the October 12, 2005, response to RAI 2.5.1.19-1 the phase bus was determined to be within the scope of license renewal as an electrical commodity group for the FRCT station. Drawing LR-BR-3000 shows the FRCT station phase bus circuits from the FRCT generators to breakers 52G-1 and 52G-2 and subsequently to breakers 52G-1N and 52G-2N.

The applicant also clarified that the phase bus (in the main generator and auxiliaries system) provides no interactions for 10 CFR 54.4(a)(2) systems. Nonsafety-related systems and components containing water, oil, or steam located in the vicinity of safety-related SSCs are included within the scope of license renewal for potential spatial interaction under 10 CFR 54.4(a)(2). The phase bus in the main generator and auxiliary systems has no water, steam, or oil pressure boundary and therefore is not within the scope of license renewal for potential spatial interaction.

In RAI 2.5.2.3-2, dated March 20, 2006, the staff noted that in LRA Section 2.5.2.3 the second bullet states: "Switchyard Bus was eliminated because none perform a license renewal intended function. Rather, transmission conductors, high voltage insulators and insulated cables and connectors perform the functions of providing offsite power to cope with and recover from regulated events."

The staff requested that the applicant address the following as to that statement:

- List (with reference to drawing LR-BR-3000) the circuits that may contain transmission conductors, high-voltage insulators, and insulated cables and connectors that provide offsite power to cope with and recover from regulated events.
- List the regulated events.

In its response dated April 18, 2006, the applicant clarified that the transmission conductors, high-voltage insulators, and insulated cables and connectors are parts of the following circuits shown on drawing LR-BR-3000:

- 34.5 kV feeds from the "B" bus of the 34.5 kV substation to transformers XMR-732-16 and XMR-732-15 via circuit breakers R144 and J69361, respectively.
- 34.5 kV feed from the 34.5 kV substation to start-up transformer SA via circuit breaker BK5.
- 34.5 kV feed from the 34.5 kV substation to startup transformer SB via circuit breaker BK6.
- 230 kV feeds from the 230 kV substation bank 9 and bank 10 circuit disconnect switches to bank 9 and bank 10 transformers, respectively. These feeds are in support of the AAC power supply credited for SBO.

The applicant stated that the transmission conductors, high-voltage insulators, and insulated cables and connectors meet 10 CFR 54.4(a)(3) because they are relied upon in safety analyses and plant evaluations to perform a function for compliance with fire protection and SBO regulations. In this discussion, "transmission conductors" refers to uninsulated high-voltage transmission cables, not the bus bar.

In RAIs 2.5.2.5-1 and 2.5.2.5-2 dated March 20, 2006, the staff noted that LRA Section 2.5.2.5.3 states that high-voltage insulators are on the circuits supplying power from the switchyard to plant buses during recovery from a SBO or fire protection event. The staff requested that the applicant describe the circuit path (which may contain the high-voltage insulators) relied upon to supply power from the switchyard to plant buses in fire protection events.

In addition, the staff noted that LRA Section 2.5.2.5.4 states that the transmission conductors are a portion of the circuits supplying power from the switchyard to plant buses during recovery from an SBO or fire protection event. The staff requested that the applicant describe the circuit path (which may contain transmission conductors) relied upon to supply power from the switchyard to plant buses in fire protection events.

In its response dated April 18, 2006, the applicant stated that offsite power from the 34.5 kV switchyard that feeds Oyster Creek station 4160V buses 1 A and 1 B, via the start-up transformers SA and SB, respectively, is credited in support of post-fire safe shutdown at Oyster Creek. Offsite power from the 34.5 kV switchyard to transformers XMR-732-15 and XMR-732-16 is credited in support of power to the redundant fire pump house. These circuits are shown on License Renewal Drawing LR-BR-3000, and are detailed as follows:

- 34.5 kV feeds from the "B" bus of the 34.5 kV substation to transformers XMR-732-16 and XMR-732-15 via circuit breakers R144 and J69361, respectively.

- 34.5 kV feed from the 34.5 kV substation to start-up transformer SA via circuit breaker BK5.

- 34.5 kV feed from the 34.5 kV substation to startup transformer SB via circuit breaker BK6.

The staff finds that the applicant's responses dated October 12, 2005, and April 18, 2006, adequately addressed the staff concerns and that the applicant did not omit any passive, long-lived components subject to an AMR in accordance with 10 CFR 54.21(a)(1). The staff's concerns described in RAIs 2.5.1.19-1, 2.5.2-1, 2.5.2.3-1, 2.5.2.3-2, 2.5.2.5-1, and 2.5.2.5-2 are resolved.

2.5.3 Conclusion

The staff reviewed the LRA, the UFSAR, and RAI responses to determine whether any SSCs within the scope of license renewal had not been identified by the applicant. No omissions were identified. In addition, the staff determined whether any components subject to an AMR had not been identified by the applicant. No omissions were identified. The staff concludes that there is reasonable assurance that the applicant has adequately identified the electrical commodity group components within the scope of license renewal, as required by 10 CFR 54.4(a), and those subject to an AMR, as required by 10 CFR 54.21(a)(1).

2.6 Conclusion for Scoping and Screening

The staff reviewed the information in LRA Section 2, "Scoping and Screening Methodology for Identifying Structures and Components Subject to Aging Management Review, and Implementation Results." The staff determined that the applicant's scoping and screening methodology is consistent with 10 CFR 54.21(a)(1) requirements and the staff's position on the treatment of safety-related and nonsafety-related SSCs within the scope of license renewal and that the SCs requiring an AMR is consistent with the requirements of 10 CFR 54.4 and 10 CFR 54.21(a)(1).

On the basis of its review, the staff concludes that the applicant has adequately identified systems and components within the scope of license renewal, as required by 10 CFR 54.4(a), and those subject to an AMR, as required by 10 CFR 54.21(a)(1).

The staff concludes that there is reasonable assurance that the activities authorized by the renewed license will continue to be conducted in accordance with the CLB, and any changes made to the CLB, in order to comply with 10 CFR 54.29(a), with the Atomic Energy Act of 1954, as amended, and with NRC regulations.

NRC FORM 335 (9-2004) NRCMD 3.7	U.S. NUCLEAR REGULATORY COMMISSION	1. REPORT NUMBER (Assigned by NRC, Add Vol., Supp., Rev., and Addendum Numbers, if any.)
	BIBLIOGRAPHIC DATA SHEET *(See instructions on the reverse)*	NUREG-1875, Vol. 1

2. TITLE AND SUBTITLE	3. DATE REPORT PUBLISHED	
Safety Evaluation Report Related to the License Renewal of the Oyster Creek Generating Station	MONTH	YEAR
	April	2007
	4. FIN OR GRANT NUMBER	

5. AUTHOR(S)	6. TYPE OF REPORT
Donnie J. Ashley	Technical
	7. PERIOD COVERED *(Inclusive Dates)*
	July 22, 2006 - March 30, 2006

8. PERFORMING ORGANIZATION - NAME AND ADDRESS *(If NRC, provide Division, Office or Region, U.S. Nuclear Regulatory Commission, and mailing address; if contractor, provide name and mailing address.)*

Division of License Renewal
Office of Nuclear Reactor Regulation
U. S. Nuclear Regulatory Commission
Washington, DC 20555-0001

9. SPONSORING ORGANIZATION - NAME AND ADDRESS *(If NRC, type "Same as above"; if contractor, provide NRC Division, Office or Region, U.S. Nuclear Regulatory Commission, and mailing address.)*

Same as above

10. SUPPLEMENTARY NOTES

11. ABSTRACT *(200 words or less)*

This safety evaluation report (SER) documents the technical review of the Oyster Creek Generating Station (OCGS) license renewal application (LRA) by the staff of the United States (US) Nuclear Regulatory Commission (NRC) (the staff). By letter dated July 22, 2005, AmerGen Energy Company, LLC submitted the LRA for OCGS in accordance with Title10, Part 54, of the Code of Federal Regulations (10CFRPart54). AmerGen Energy Company, LLC requests renewal of the operating license for OCGS (Facility Operating License Number DPR-16), for a period of 20 years beyond the current expiration date of midnight April 9, 2009.

OCGS is located in Lacey Township, Ocean County, New Jersey, approximately two miles south of the community of Forked River, two miles inland from the shore of Barnegat Bay, and nine miles south of Toms River, New Jersey. The NRC issued the OCGS construction permit on December 15, 1964, the OCGS provisional operating license on April 9, 1969, and the OCGS operating license on July 2,1991. OCGS is a single unit facility with a single-cycle, forced-circulation boiling water reactor (BWR)-2 and a Mark 1 containment. The nuclear steam supply system was furnished by General Electric and the balance of the plant was originally designed and constructed by Burns&Roe. OCGS licensed power output is 1930 megawatt thermal with a gross electrical output of approximately 619 megawatt electric.

This SER presents the status of the staff's review of information submitted through February 15, 2007, the cutoff date for consideration in the SER. The staff identified open items that were resolved before the staff made a final determination on the application. SER Section1.5 summarizes these items and their resolution. Section 6.0 provides the staff's final conclusion on the review of the OCGS LRA.

12. KEY WORDS/DESCRIPTORS *(List words or phrases that will assist researchers in locating the report.)*	13. AVAILABILITY STATEMENT
10 CFR 54, license renewal, Oyster Creek, scoping and screening, aging management, time-limited aging analysis, TLAA, safety evaluation report	unlimited
	14. SECURITY CLASSIFICATION
	(This Page) unclassified
	(This Report) unclassified
	15. NUMBER OF PAGES
	16. PRICE

NRC FORM 335 (9-2004)   PRINTED ON RECYCLED PAPER

www.ingramcontent.com/pod-product-compliance
Lightning Source LLC
Chambersburg PA
CBHW051454170526
45166CB00001B/241

* 9 7 8 1 5 0 0 1 6 4 8 0 5 *